Saser Kangri III (7495 m, N face) from Camp 1 (6000 m) on the North Shupka Kunchang glacier (Divyesh Muni)

THE HIMALAYAN JOURNAL

Honorary Editors

RAJESH GADGIL
NANDINI PURANDARE

VOLUME 69
2013 - 2014

Published for

THE HIMALAYAN CLUB

OXFORD UNIVERSITY PRESS
DELHI LONDON NEW YORK
2014

THE HIMALAYAN JOURNAL

(Published since 1929)

To encourage and assist Himalayan travel and exploration,
and to extend knowledge of the Himalaya
and adjoining mountain ranges
through science, art, literature and sport.

ISBN
0199456127

ISBN
9780199456123

THE HIMALAYAN CLUB CENTRE

314, Turf Estate, Shakti Mills Lane
Off E. Moses Road, Mahalaxmi, Mumbai 400 011, India
E-mail: info@himalayanclub.org
Website: www.himalayanclub.org
Phone (91-22) 2491 2829

E-mails

Club Office	:	info@himalayanclub.org
Hon. Secretary	:	secretary@himalayanclub.org
Hon. Editor	:	editor@himalayanclub.org

CONTENTS

Cover: Sari lake and Jot. Himachal Pradesh (Harish Kapadia)

Back Cover: Climbing Shiva peak, Pangi valley, Himachal Pradesh. (Mick Fowler)

Frontispiece: Saser Kangri III (7495 m, N face) from Camp 1 (6000 m) on the North Shupka Kunchang glacier (Divyesh Muni)

Title Page: Climbers moving ahead of Camp 2, with the west ridge towering behind. (Brig A. Abbey)

Background Contents pages: Nyainbo Yuze N face and Ximen Co (Tom Nakamura)

EDITORIAL

Time has a tendency to slip by in a way that is so elusive, particularly, when one is trying to tie up all the events that have happened in the world of the mountains since the last issue. Was the disaster that hit Uttaranchal covered already? Did we talk about the passing of Nawang Topgay Sherpa in the last issue? Or was it covered in the *E letter* and not in the *HJ?* – scouring notes, dates, past volumes, scribbling in notebooks when ideas occur – all is a maze when one sits down to round it all up in the first ever editorial that one has been forced to pen!

The only way to do it I realise is to do it. So here goes.

Our sincere apologies again, for this late late issue. *The Himalayan Journal* that links Himalayan Club to the climbing world is faltering. Changing times, increasing pressures and reducing hands-on-deck volunteers, have all contributed to this situation despite good intentions. We are in the process of a serious rethink, with an SOP being put in place to help us adhere to all timelines and promises. Bear with us, please. I must mention that the old war horse, the solid Harish Kapadia, now Editor Emeritus, *The Himalayan Journal* was roped in again this year to help put together bits and pieces of this volume. Rajesh Gadgil, of course worked tirelessly on most of *HJ 69* but then he is a mountaineer and needed to go climb.

The one aspect that all of us can continue to take pride in is the quality and content. This issue covers a range of interesting writings ranging from a study of *HJ* covers over the ages to a roundup of mountain festivals of the world to the description of a condition called glacial lassitude. Explorations by Himalayan Club members like that of the Jadh Ganga valley; the trail to the historic Chaukan pass leading to Burma; the first crossing of the Chaukhamba col make wonderful reading as do the significant climbs such as the first ascent of Plateau Peak by Pradeep Sahoo's HC team, the amazing feats in Kishtwar by Mick Fowler and the continuing story of exploratory climbs in the East Karakoram by Divyesh Muni.

Talking about expeditions, one more is significant enough to mention here. On the night of 15 – 16 June 2013, a cloud burst in the Kedarnath area brought down huge avalanches that crashed into the Chorabari lake. The banks of this lake burst and huge cascades came crashing towards the Kedarnath temple, sweeping away everything in its wake – the devastation was massive, even lower down – villages, pilgrims, properties, fields, roads vanished. That the disaster was manmade is an academic and valid argument but at the time of crisis it was rescue and rehabilitation that was the need of the hour. An Indian army expedition led by HC President Brig. Ashok Abbey, fresh after a successful ascent of Trisul west ridge, had just returned to Joshimath. As most of the team members were former instructors and students of the High Altitude Warfare School, they immediately joined the rescue operations, as a result of which hundreds of stranded civilians were rescued.

In May 2014, Sherpa Tenzing Norgay would have been a 100 years old. In an ironic twist, this year, his beloved Everest threw up the worst accident ever on a single day, when 16 route opening Sherpas were killed in an avalanche on 18 April, 2014. All climbing activity on the mountain ceased as debate on compensation, safety, ethics etc have been raging all over the climbing world yet again. The ongoing oral history project on the 'Climbing Sherpas of Darjeeling' somehow seems more significant as its effort is to give a human face to these brave carriers of yore by telling their stories.

Aspi Moddie was one of the Himalayan Club's oldest members (since 1949); he was its first real Indian President, a man who loved and nurtured the Club; a man with a vision that has held it together in its current form. In the words of Meher Mehta – '…at the time of his death at the age of 92, he was the eldest, or near to that, the longest and most respected serving member of the Club, covering a period of 65 years.' In his passing the Club has truly lost its friend, philosopher and guide. Sadly, other friends have passed on too – Richard Martin Scott, Ian McNaught-Davis and 'Guruji' Nawang Topgay Sherpa.

Since the time of Volume 68, the Himalayan Club achieved yet another milestone – it concluded its first Strategic Review and has begun implementation of short, medium and long term reforms.

And so I sit here, continuing to check and recheck notes, thinking how I will be able to fit into the shoes of our predecessors and do justice to the great institution of *The Himalayan Journal.*

Nandini Purandare
Co-Editor

1

The Story of the Climbing Sherpas of Darjeeling – first impressions

Deepa Balsavar and Nandini Purandare

This is no ordinary morning. For the Toong Soong community, this chilly November day marks a poignant farewell.

Today is the funeral of Nawang Topgay Sherpa, the last surviving Tiger Medallist of Darjeeling. Descending steep stairs and winding alleys, passing buildings and shanties, we finally reach lower Toong Soong that clings precariously to the hillside midway between upper Toong Soong and the valley floor. Guruji (Respected teacher), as Topgay was known, died quietly in his sleep yesterday after lunch, while watching TV with his grandson.

We enter his home, looking as sombre as possible and see people in every room, eating, drinking Pepsi and chang, gossiping and catching up with each other. Our friend Phurtemba comes forward with a wreath, hand crafted by him that he gives to us to place on the crate-like coffin. Following their Buddhist beliefs, Topgay's community of Sherpas is sending him to his resting place without signs of overt lament. His body is deliberately curled in foetal pose, exiting the world the way he entered it.

Nawang Topgay Sherpa. (courtesy Nawang Topgay collection)

Born in the Tibetan year of the Ram, 82-year-old Topgay lived for the better part of his life in Toong Soong, until his passing away on 2 November, 2012.

Young Topgay had migrated in search of work to Darjeeling when he was a lad of 14. He was taken under the care and guidance of his father's younger brother, Tenzing Norgay and lived with

The funeral procession of Topgay Sherpa.

him till he found a place of his own. It was with Tenzing that he began his expedition work as a high altitude Sherpa. He was one of the youngest porters to reach South Col on the 1953 Everest expedition and soon after when the Himalayan Mountaineering Institute was established he was appointed instructor. He later went on to teach at the Nehru Institute of Mountaineering in Uttarkashi. Known popularly to scores of students as Guruji, he retired from NIM and returned to Darjeeling where he spent the rest of his life.

And so we stand on the hillside in the beautiful morning sun, waiting for the coffin with the lone marigold wreath on it and the funeral procession to wind its way up and up, back to the upper section of the basti and towards the Sherpa cemetery at the southern end of Toong Soong.

'We' are the Sherpa Project researchers, spending time in Darjeeling, collecting and collating oral histories from the families of the great climbing Sherpas of Darjeeling. Who were these men? Where did they come from? What did they do?

This is the task that we set for ourselves six months earlier when we attended the first death anniversary ceremony of Topgay's cousin and another great Tiger, Nawang Gombu. He too was a nephew of Tenzing – his mother was Tenzing's sister – and he was also on the 1953 Everest expedition that put Tenzing and Hillary on top of Everest. Gombu also belonged to Darjeeling but the future he had secured for his children was in a world far removed from Toong Soong.

As our stays in Darjeeling lengthened, we came to hear and understand more about a remarkable community of people; of men and women who, entering India as illiterate, impoverished migrants had been able to carve a niche for themselves in this country and a name for themselves in the world. Whether upwardly mobile like the families of Gombu and a few other climbers or humble like most of those who live in Toong Soong, the stories literally and figuratively reveal nothing less than a vertical ascent.

The beginnings

To introduce the community we are talking about, a little history is in order.

If you ask an old Sherpa or Sherpani from Darjeeling where he or she comes from, the answer is usually 'Solu khumbu' or 'Shar khumbu'. Solu and Khumbu divided by a third stretch known as Pharak are actually a series of valleys in the vicinity of Mt. Everest in Nepal. The word Sherpa derives from 'Shar wa' ('Shar' means 'east' and 'wa' or 'pa' means 'people') and refers to inhabitants of eastern Nepal who are of Tibetan origin. It is believed that the Sherpa ancestors fled from Kham in Eastern Tibet in the 1600s to settle in the upper Khumbu valleys of Nepal.

At 3500 – 4000 m, Khumbu, where most of the climbing Sherpas came from, is a hard environment to live in. The villages of Thame, Namche Bazaar, Khumjung and Kunde lie scattered in the playground of the highest peaks in the world. In the early days the people were agriculturists (growing barley and potatoes), herders (grazing yaks and naks in the higher pastures) and traders (trading with Tibet across the high passes). Gradually many Sherpas moved to the lower valleys to grow better crops and enjoy a less extreme environment – to Solu in the south. Dorjee Lhatoo, a distinguished climber and our 'go-to' person for all queries on Sherpas has this to say about the inhabitants of the two areas 'Solu Sherpas call Sherpas of Khumbu "snot eaters" and the Sherpas of Khumbu call their Solu counterparts "crawlers under the bridge".' (This is a private joke among the two groups. Khumbu Sherpas are unfamiliar with fishing so they say Solu Sherpas crawl under the bridge for worms. The Solu Sherpas respond by pointing to the permanently snotty noses of the children of colder Khumbu).

So when did the Sherpas start coming to Darjeeling and how did they get here? In an article he wrote for the *HJ* Vol 35, 1976 - 78, Lhatoo noted:

> There is no recorded history of the first arrival of our people in Darjeeling. But there is evidence of their being here as early as towards the end of the 19th century. Like many of the plainsmen who came up as the followers of the British, and the eastern Nepalese who came across attracted by labour and other employment prospects in the tea plantations, many Tibetans and Sherpas came to work as 'coolies' and 'dandy bearers' for the tourists after Darjeeling was established as a hill resort by the British Government.

The first migrants walked uncharted lands, following the sun and asking directions at infrequent settlements. It must have been an arduous journey in any season. We have this account of the trip made by a Sherpa when he was a boy of six or seven from his home town of Yatung in the Chumbi valley of Tibet to Darjeeling:

'I came walking I remember. I came walking over the Nathu la. It must have been in October – November. I remember the time because it was orange season. Now I believe you can descend from Nathu la into Chumbi valley in one hour and in two-three hours you are at Yatung. But I remember my mother saying that we walked for nine days. Nine days from Yatung (Tibet) to Bhutia Basti in Darjeeling.

'I remember Nathu la vividly. Then it was wilderness. Nothing! Complete wilderness. We were the only people crossing the pass. My uncle - my mother's youngest brother - and his wife and friend came to Yatung with six horses to fetch us. My older sister and younger brother were on one horse and on the other horse were my mother and the baby. The rest of the horses had the provisions and whatever my mother could bring.

'And I walked with the three elders all the way. My mother used to flatter me and these two men would also say "He's a strong boy; he doesn't like to ride a horse." So I didn't want to ride a horse because of my ego. But I used to be so tired I slept while walking I'm told.'

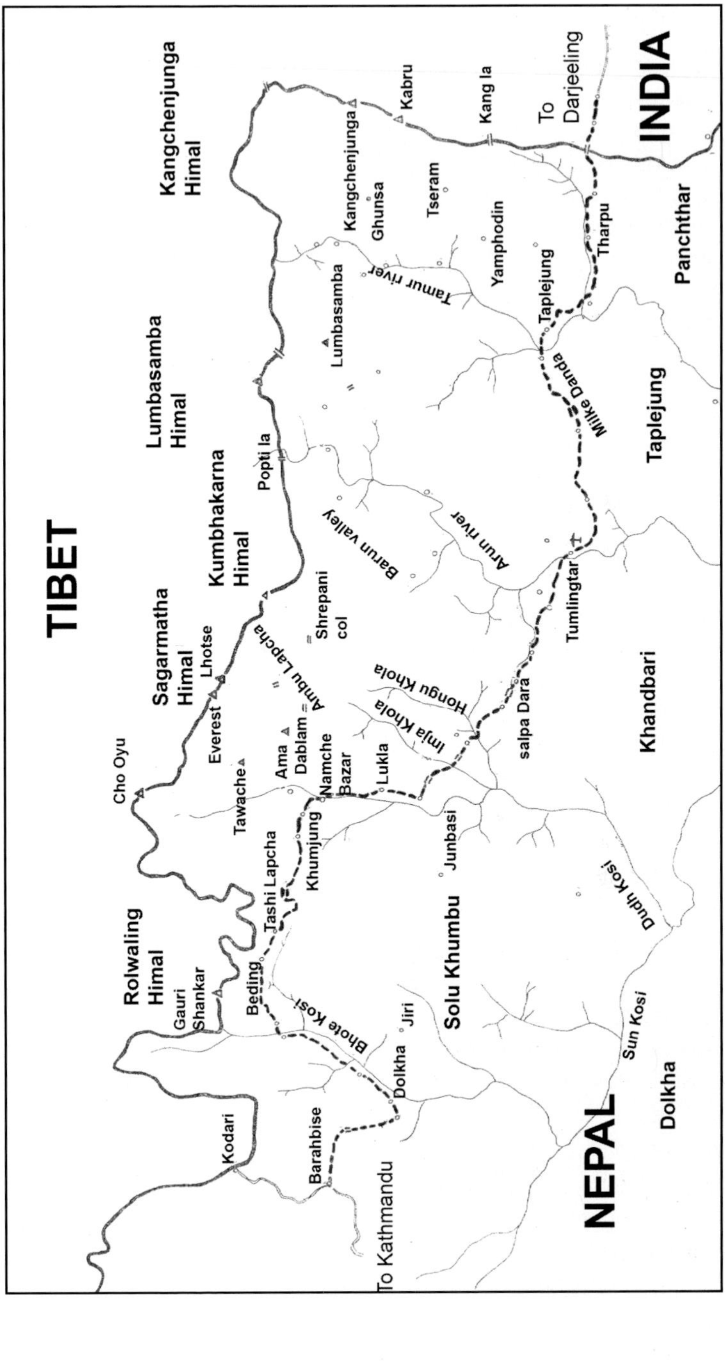

One of the routes taken by Sherpas from Solu Khumbu to Darjeeling.

Toong Soong.

All the early immigrants from Nepal and Tibet to Darjeeling would have made a similar journey. Another older Sherpa shared this with us: 'How did they find the way? They looked at the early morning sun. They saw from where it rose. They knew they had to go east so they would go in the direction of the sunrise. From Nepal they came walking for 15 days. You walk from sunrise to sunset and then you reach a village. You could request for a place to stay.' There was trust in those days and a place by the hearth and a glass of chang for every weary traveller.

If you had to follow one of those trails from Nepal today, you would walk for about 375 km, cross seven passes ranging from 2300 m to 5800 m, and cross the watersheds of major river valleys such as Bhote Kosi and Dudh Kosi. On an average, you would climb and descend about 5000 m to cross one river valley.

The new and promised land

And so they came to settle in Toong Soong, a hill side in Darjeeling. It is the only side with no views of the snowy mountains. There are nicer parts of course but these were taken by the British sahibs who built the hill station. Toong Soong is a buzzing shanty town. If you are on Chowrasta in Darjeeling, look south. There is a street on the left, lined with open food stalls, vendors selling vegetables, flowers, meat and fruit. Past that, there is a crowded township spanning the hillside with a narrow main artery slicing through it. Along this road is the Sherpa gompa - the centre of religious, social and cultural activities for the community. It is a comfortable cheery temple with men and women volunteers maintaining the place and always trying to build one more floor! Above, below and alongside the temple are apartments vying with houses, vying with chawls, vying with budget hotels for a piece of the sun. And when the sun is out, the dice is rolled by the young men, dogs and cats stretch, stuffed toys are washed and hung out to dry and women knit or wash or gossip along this lively street.

When the first Sherpas built their wooden homes in Toong Soong though, the hillsides were thickly forested. There was plenty of water and firewood to chop and sell for money. Nima Tsering, whose father was Gyalzen Sirdar and who lives in one of the oldest houses in Toong Soong told us that after sunset it was too dangerous to venture out. Even Mall road, now the busiest thoroughfare in Darjeeling, was a desolate area, and once the sun had gone down, the children remained indoors huddled in fear.

This is Dorjee Lhatoo again on the origin of the name Toong Soong:

> I heard from the early Sherpas that Toong Soong is a corruption of Thung Soom. Thung is shankh (shell or conch). There were three prominent rock features on this road. One is still there. From Chong Rinzing's house if you look up – there's a big rock. We used to do our rappelling on that. That looks like a shell. There was one close to the Sherpa gompa. And there was one at the bifurcation, where the three roads meet. They have blown that up and built a big house there. *Soom* in Tibetan is three. Thung Soom – Three shells – one is still there. You can't miss it. The old Sherpas still say Thung Soom.

Today, we are visiting the neat cottage that Ang Tshering built.

Even steps lead down to a broad paved terrace overlooking lower Toong Soong and the valley. We have seen pictures of Ang Tshering sitting here whittling a piece of wood and even now, ten years after his death his presence is everywhere. The entrance to his spotless green painted house bears a small black sign 'Ang Tshering Sherpa, H. No. 10/1, Ganesh Gram, Toong Soong

Four of the seven daughters of Ang Tsering with their cousin Nima Tsering, son of Ajeeba Sherpa.

Basty, Darjeeling.' Many climbers especially foreigners come actively seeking this place for Ang Tshering is a legend. He was awarded his Tiger medal for his role in the 1952 Everest expedition and he continued climbing well into his sixties. By the time Ang Tshering passed away just months short of his hundredth birthday, he had amassed medals and citations from all over the world and these are now part of a loving legacy guarded by his daughters.

Ang Tshering's daughters, Pema Diki, Phur Diki and Lakpa Chamji live in this house now. 'Altogether we were 12 brothers and sisters – five boys and seven girls' explains Pema Diki. Only five sisters and the brother Tharchen are still alive. There is a lot of laughter in this house. Hot tea, momos and Hindi film dialogues are dispensed with equal generosity. When they talk about their father however, the voices lower and soften and they interrupt each other frequently to make sure that not the tiniest detail is overlooked. Ang Tshering's medals are mounted on blue velvet and encased in a glass cabinet.

Curiously, the women do not know much about their father's expeditions except for a shaky account of the infamous German Nanga Parbat one in which four Germans and six Sherpas lost their lives. Phur Diki easily

1952 Pre-Everest. Standing (l to r): Pasang, Kirken, Nauri, Ang Dawa Phurkey, Jombe, Pasang Phutar Jockey. Standing, middle row (starting from the middle l to r): Topgay, unidentified, Tenzing, Da Thendup, unidentified, Pemba Sungdar. Seated (l to r): Nawang Gombu, Ang Balu, Ang Temba (above), Da Namgyal (below), Da Thondup, Ang Nyima, Phu Tchering (Courtesy Dorjee Lhatoo collection)

the actor in the family begins the account. She punctuates her narrative with actions and voice changes interrupted occasionally by the older Pema Diki who mutters loud stage whispers behind her.

'In 1934 Babuji (father) went to Nanga Parbat – we have only heard the story – we are not sure. It was a German expedition. There was a school teacher – Merkl – and the Sherpas: Gaylay and Dakshi. I don't remember the others. Babuji, Gaylay and others were sent up by the leaders. When they went up, the weather turned bad. At base camp everything was good. They had taken food up but it ran out while they waited for the weather to clear. The people below could not take food to those who were stranded higher up. They had reached Camp 8. What to do? Babuji thought we can't go up, can't go down... Seven days had passed. School master was very sick and weak. Gaylay stayed with Merkl who died that night. Before he died Merkl asked Babuji to get help. On the way down Babuji became snow blind. He could not see and there was nothing to eat. He finally came down eating ice. Nine days he survived on ice. So many died. Some bodies were still on the ropes... Finally he could see the base camp. 'Hai hai' – he tried to call but no sound would come. The people could not see or hear but he could see them moving around. He went further down. One man who was taking food from one tent to another finally noticed him. They thought they were seeing a ghost … someone helped bring him down. "Where is Merkl?" they asked. "They are coming down slowly. There is no food, send some people". They sent five people up. Babuji was brought to base camp and then admitted to hospital. His feet were frost bitten and they put leeches to stem the gangrene. He lost some toes.'

Starting from the bottom

The first real record of the presence of Sherpas in mountaineering was noted in 1907 when Dr. A.M. Kellas took high-altitude porters with him from Darjeeling to the Sikkim Himalaya. Ever since then, Sherpas have been mentioned in almost every Himalayan mountaineering book. Early photographs show them as small smiling men with cropped hair or pigtails, wearing happy combinations of Tibetan and Western clothes. They carried loads, made camps and cooked. They were referred to as coolies or porters or load bearers, sometimes with affection but more often in a patronising manner. Their employers, the Europeans who climbed and explored and wrote books during the early 1900s mention their childlike laughter and fondness for pranks, their love of chang and occasional petulance and sulks. While there was often genuine fondness for individual porters, the Sherpas on the whole were regarded as nameless pieces of essential equipment necessary to allow the sahibs to get to the top. And yet these were individuals who risked their lives on an annual basis, not for the lure of the mountains but in order to earn a living.

The story of the Sherpas – from cultivators to load carriers to faithful climbing companions – makes a riveting saga. Europeans eyeing the mighty 6000+ m peaks of the Himalaya noticed the Sherpas were naturals at learning mountaineering techniques. They had one other quality besides – adaptability. Eric Shipton wrote in 1937 during survey work in the Nanda Devi region that two of the local Lata porters they had hired 'worked splendidly and with a little training would be as good as the Sherpas. But, they have one tremendous disadvantage and that is that their religion forbids them to eat either with Europeans or anything cooked or touched by Europeans or Indians of other castes. When a party is engaged in a long and difficult task, this taboo would produce an impossible situation. With the Sherpas I am in the habit of eating out of the same dish and drinking out of the same mug and no one loses caste or feels embarrassed. Later in the year, when we were employing some Dotial porters and ran short of food, the Dotials who had finished their own food, allowed themselves to become feeble with hunger, rather than eat the rice which we had been carrying in our rucksacks. Ang Tharkay always becomes infuriated by this prejudice and taunts the victim mercilessly'.

Hard way up

At the end of the 19th century the Himalaya of the Raj was the centre of adventure; an intriguing and exciting destination for the British, the French, the Swiss and the Germans.

Parties from all these places came in the first decades of the new century to climb, and the hills were alive with the sounds of hundreds of porters hauling luggage and equipment. It is no accident that the words used for climbing include assault, attack and conquer, for the attitude of the westerners was that of an army launching an offensive against a worthy opponent. Early expeditions were massive affairs. Porters carried equipment, tents and food as also luxuries like cots, tables, silver tea services, linen napkins and tin baths for the sahibs to wash off the dirt of the mountains! This may not seem too much of a hardship until you realise that even base camp in the Himalaya is often higher than most mountains in the Alps. In 1931, during the Kangchenjunga expedition, the team comprised five Germans, three English, two Swiss and one Austrian; a total of 11 members. 350 coolie loads were assembled. Prof. G. O. Dyhrenfurth, the leader, stated that the load was not excessive for the number of members on the team. Rationalising, he wrote 'The Vissars for instance, took on their last expedition 450 porters for half the number of Europeans. For the baggage of the 3rd Everest expedition (containing the same number of Europeans) 770 coolie loads were required (although this was divided over 70 porters and 350 animals)'. Dyhrenfurth's expedition provided only 70 pairs of shoes so many coolies climbed barefoot above 4000 m – the height where the blanket of snow starts.

Let's have a little visual imagery: imagine yourself in a plane and looking out the window. Cruising altitude for most planes is about 10,500 m. At 8848 m the summit of Mt. Everest is less than 1700 m below you and Kangchenjunga at 8586 m is just 300 m lower. Now imagine you are not in the nicely pressurized cabin but on the slope, contending with -20 to -35 degrees Celsius temperatures, howling winds, icefalls, and the dangers of avalanches and snow blindness, all with a load on your back and clothing that is ill-fitting and often insufficient. Part of the weight borne by you would be bottled oxygen for the members. This was brought into use as early as 1922. One bottle at that time contained approximately 240 litres of oxygen and four bottles were fixed on a porting fixture. With the additional elements this meant a weight of about 14.5 kg. This bottled oxygen was used only after gaining a height of 7000 m. This means someone

Joan Townend, Secretary of the Himalayan Club, Darjeeling. (courtesy Ang Tsering collection)

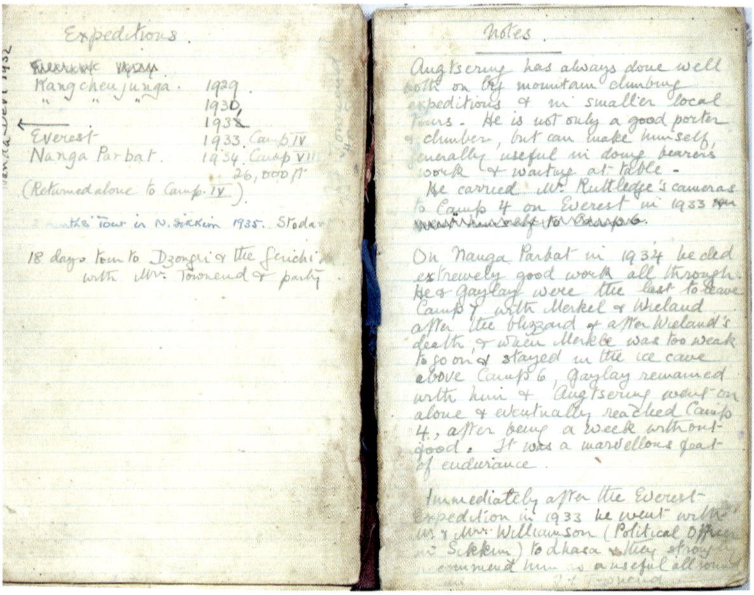
Expeditions.
Kangchenjunga. 1929.
" " " 1930,
1938
Everest 1933. Camp IV
Nanga Parbat. 1934. Camp VII 26,000 ft.
(Returned alone to Camp IV).
months' Tour in N. Sikkim 1935. Stoda
18 days tour to Dzongri & the Guicha with Mr. Townend & party

Notes.
Ang Tsering has always done well both on big mountain climbing expeditions & in smaller local tours. He is not only a good porter & climber, but can make himself generally useful in doing bearer's work & waiting at table.
He carried Mr. Ruttledge's cameras to Camp 4 on Everest in 1933

On Nanga Parbat in 1934 he did extremely good work all through. He & Gaylay were the last to leave Camp 7 with Merkel & Wieland after the blizzard & after Wieland's death, & when Merkel was too weak to go on & stayed in the ice cave above Camp 6, Gaylay remained with him & Ang Tsering went on alone & eventually reached Camp 4, after being a week without food. It was a marvellous feat of endurance.

Immediately after the Everest Expedition in 1933 he went with Mr & Mrs Williamson (Political Officer in Sikkim) to Lhasa & they strongly recommend him as a useful all round

Pages from the Himalayan Club issued 'chit book'.

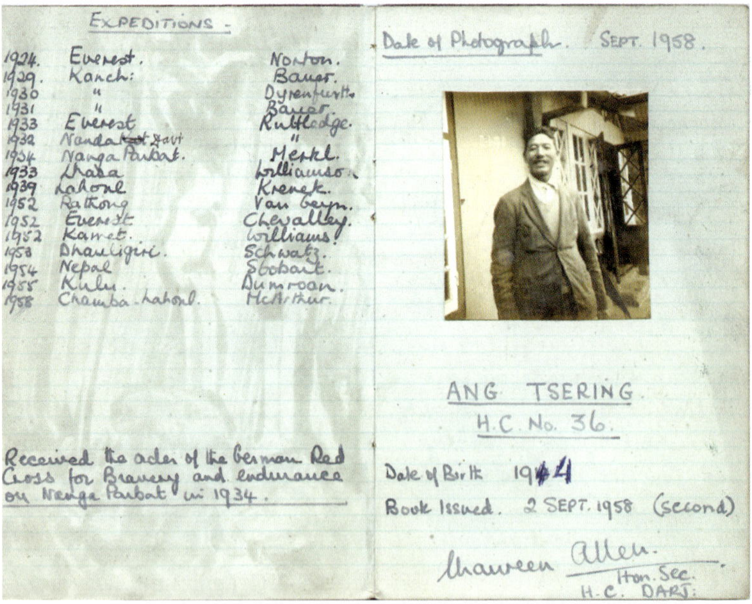
EXPEDITIONS -

1924.	Everest.	Norton.
1929.	Kanch:	Bauer.
1930	"	Dyrenfurth.
1931	"	Bauer.
1933	Everest	Ruttledge.
1932	Nanda Devi	"
1934	Nanga Parbat.	Merkl.
1933	Lhasa	Williamson.
1939	Lahoul	Krenek.
1952	Rathong	Van Gyn.
1952	Everest	Chevalley.
1952	Kamet.	Williams.
1953	Dhaulagiri.	Schwatz.
1954	Nepal	Stobart.
1955	Kulu.	Dumroan.
1958	Chamba-Lahoul.	McArthur.

Received the order of the German Red Cross for Bravery and endurance on Nanga Parbat in 1934.

Date of Photograph. SEPT. 1958.

ANG TSERING
H.C. No. 36.

Date of Birth 1904
Book Issued. 2 SEPT. 1958 (second)

Maureen Allen.
Hon. Sec.
H.C. DARJ:

Ang Tsering's chit book.

would have to carry it at least up to this altitude. Each bottle lasted about two hours and all the oxygen would be used up after a maximum of eight hours of climbing. The Tibetan and Nepalese porters nicknamed these oxygen bottles 'English air'.

The Himalayan Club Tiger medal.

By the 1920s, Darjeeling had emerged as the base for major intrepid Himalayan reconnaissance and Everest was the dream everyone wanted to capture. The climbing world is littered with names made famous through the expeditions of the period: Gen. Charles Bruce, Dr. A.M. Kellas, Tom Longstaff, Frank Smythe, Eric Shipton, H.W. Tilman, and the Duke of the Abruzzi … Edward Norton, Dr. Somerwell, George Mallory and Sandy Irving ... the list goes on. They all came to Darjeeling to hire their porters and attempt climbing Everest from the Tibetan side as Nepal was closed to foreigners. For a period though, this route was also closed down. In 1924, on the return journey from Everest, one of the members of the expedition broke away from the normal route to visit Tsangpo - a forbidden trip - and later on in London, a photographer arranged for some Tibetan lamas to dance at lectures to give them publicity. The authorities at Lhasa were very annoyed by both occurrences and put a ban on further expeditions, which was not lifted until 1933.

It is against this backdrop that in 1927 people in Calcutta and Simla began to think of founding a club to encourage people to visit the mountain regions on India's northern borders. The Himalayan Club was born in 1928 and its most active cell was Darjeeling where all the action was. The Club took the responsibility of organising Sirdars and porters and a strong bond between the community and the club took root; a relationship that continues even today.

In 1934, Joan Townend, became Hon. Secretary of the Himalayan Club's Eastern Section. She was deeply distressed by the loss of Sherpa lives on Nanga Parbat and keen that individual Sherpas should be suitably commemorated (Ang Tshering - whose descent from the mountain was told to us by his daughters - was on this expedition), but she found that details of their careers were lacking. To facilitate the collection of information, Mrs. Townend suggested that each Sherpa should be issued a 'chit book' containing his photograph and wrapped up in a mackintosh case. Leaders of expeditions were expected to fill in details at the end of each expedition and remark on the services of the porters. General Bruce arrived in Calcutta in December of the same year and went up to Darjeeling to give a feast to all his old friends and distribute this prized possession. Indeed we are lucky that we have come across many of these, lovingly preserved by the families but with pages that probably contained unsavoury comments assiduously torn away!

When men become Tigers

The Tiger Medal was instituted by the Himalayan Club in 1939 - an honour bestowed on a handful of the worthiest Sherpas to recognise outstanding service rendered on British expeditions. Though 'Tiger' was a term George Mallory used in the 1920s, the medal was announced almost two decades after, for select Sherpas who boasted exceptional feats of climbing and bravery on hostile ice and rock. The very first medals were given out to ten handpicked Sherpas on 30 May, 1939.

The *HJ* Vol XII (1940) includes a note by Joan Townend on how Tiger Badges for Sherpas had been decided on and the Himalayan Club Committee was to take great pains in the years to come to ensure that the right Sherpas got the awards.

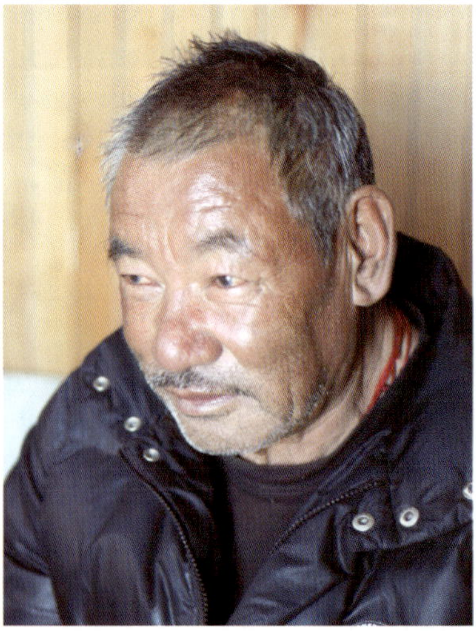

Sona Sherpa, the last Tiger in Darjeeling.

The note reads thus:

The Grading of Sherpa and Bhotia Porters

At a Committee meeting of the Eastern Section of the Himalayan Club on the 6th of February 1939 it was decided to create a superior grade for experienced climbing porters, and to give them 8 annas a day extra pay beyond the rate paid to others, for work above the snow-line[1]. It was first suggested to give them a smaller increase at any time, until it was realized that this might prevent travellers from engaging them for ordinary treks, where climbing skill and experience are not so valuable; and since they count on employment on these journeys to keep them between the high climbing expeditions, it was decided to give them a larger increase when their greater skill is being used.

It was suggested that these men should be given the name of 'Tigers', together with a badge representing a tiger's head. There were a number of criticisms of this name, but none of the alternatives appeared suitable. 'Climbers' is a term already used for Europeans of the party; 'guides' would give a false impression, for it is most undesirable that the porters should be looked upon as guides in the Swiss sense; and since the name 'Tiger' has been fairly constantly used since the Mount Everest expedition of 1924 for the picked porters who have gone high, it has been adopted as the best name put forward.

Striking Gold

We had believed that after the death of Nawang Gombu, Nawang Topgay had been the last Tiger in Darjeeling; but our readings and interviews had not answered the question: 'When was the last Tiger badge awarded?' Without knowing the answer to that, we couldn't be sure that there were no other living Tigers. Then one rainy morning in our little cottage in Singmari, pouring over past issues of the Himalayan Journal, we have a Eureka moment. Tucked away in an issue from 1967 we find an exhaustive list of Tiger Badge recipients. Towards the end of the list is an entry – 'Sona (Darjeeling), Everest 1965'. Feverish searches of later issues produce no further lists. These seem to have been the last Tiger medals given out. Is Sona then an unknown Tiger? Would he still be alive? Fingers crossed, we set out to locate him.

'Sona Sherpa? There is no such Tiger medallist,' respond local contacts. 'Wait a minute, an alcoholic who goes by the nickname 'King Kong' walks to Chowrasta from Toong Soong every day.'

1. To give a perspective of the salaries and compensation, in 1931, a Sherpa was paid one quarter of a rupee a day, plus five rupees per joint lost by frostbite.

'Yes, he's the one!' Dorjee Lhatoo had already told us that the climber Sona used to be better known by that name. 'He cannot be the medallist you are looking for, but his name is Sona. He lives with his daughter.' Wanting to check for ourselves, we ask Phurtemba to organise a meeting with Sona in Toong Soong. 'Tomorrow morning', he says. 'That way I can warn his daughter to keep him sober'.

The next morning we take our usual route. Just past the Sherpa gompa, we see a shaggy, bumbling grown man being gently lowered onto a stool and given a thorough washing by a younger woman – anything is possible in Toong Soong! We settle down in Phurtemba's little home to wait and a few minutes later the same timid bear of a man enters. Well scrubbed now with water glistening in his wet hair, his daughter has kept her promise and delivered a sober Sona for the occasion. A tremor visibly shakes his hand as he dabs a crumpled handkerchief on his furrowed forehead. This big and gentle but wasted soul with heartbreaking rheumy eyes was bestowed the Tiger for his role in the first successful Indian Everest expedition of 1965. When Captain M S Kohli placed nine Indians on the summit, Sona was one of the support staff. But his recollection of those days is dim. He does not know where his 'prized' medal is, probably in his hut in Gwaldam where he worked as a chowkidar of some sort ... 'When my name was called for a medal, I thought, there is nothing special I have done, how did I get this? So I was really left very surprised and wondering.' At this point Phurtemba adds 'May I say something? Mr. Sona says that he had nothing ... But he had a lot! He was young, he had energy, he had strength – which is why everyone favoured him and took him on expeditions. That is what he had, so much of, even though he says he had nothing.' 'You are right' smiles Sona, 'But I have nothing now; I have no support in life. When I drink, that is some *sahara* (solace).'

Twists and turns

After Tenzing's success on Everest an exuberant Nehru started the Himalayan Mountaineering Institute to create 1000 more Tenzings. Scores of youngsters fired with enthusiasm came to train but, after living in seclusion for decades, Nepal finally threw open its doors to foreigners and the wave of expeditions and jobs slowly and inexorably flowed back to where it all began, back to Solu Khumbu. The wise

(L to R) - Da Namgyal, Harish Kapadia, Gyalzen Mikchen, Ang Temba and Dorjee Lhatoo at Chowrasta, Darjeeling. (Harish Kapadia)

Sherpas who invested in their children's education are the ones who weathered the storm, but many others floundered.

Harish Kapadia, a household name among trekkers and Sherpas alike, shares a story from around 1970 about the gentle Tiger, Da Namgyal. Shy but extraordinarily strong; he was one of only three Sherpas to press upwards of South Col on Everest in 1953. Walking with Lhatoo in Darjeeling bazaar, Kapadia saw Da Namgyal and Gyalzen Mikchen, another celebrated Sherpa, selling woollen sweaters at a roadside stall. 'I asked what they made per month. They mentioned 200-300 rupees. As Class 4 employees of the Bengal government they got no pension.' Kapadia published a picture of all three Sherpas in the Himalayan Journal and a couple of British magazines. 'It caused a furore. My caption under the photo read: "Two sweater sellers and one future sweater seller of Darjeeling".'

It's growing late and our days in Darjeeling are coming to a close, for this trip at least. But there is always time for another story and this time it is about the women, at last.

A Sherpa wife pining for her husband and unable to put up any longer with the fear and worry when he was away, sent word to him at

The magnificent men. Standing (l to r): Da Thondup, Da Thendup, Da Namgyal, Tenzing Norgay, Ang Nyima, unidentified, Pasang Dawa. Sitting (l to r): unidentified, Nawang Topgay, Ang Temba. (courtesy Nawang Topgay collection)

base camp that she was ill and needed hospitalisation. The husband immediately resigned his job and set off for home in double quick time making the journey of five days in two. Panting for breath, he threw open the door of his house to find his wife calmly cooking in the kitchen. He was livid with anger and berated her for the fright that propelled his desperate journey home. The wife listened and then replied; 'Now you know what I go through every time you leave'. Her husband never said a word of reproach to her again.

Climbing mountains is an uncertain business and Sherpa women are equal partners with their men when it comes to earning a living and keeping the family going. Indeed many of them were load carriers themselves, carrying the coin boxes filled with porter's wages up to base camp. Ang Tharkay's wife was a road contractor and many others started independent businesses and small shops. In social situations we hear the banter and even mild flirtation between the sexes and see the confidence with which Sherpa women move about.

There is so much that we have begun to learn: of the rituals and superstitions, family intrigues and alcoholism; ambitions and disappointments and there are questions that remain scrawled in the margins of our note books: Where are the papers of the Sherpa Association started by Tenzing? Can the Sherpas of today consider jobs in HMI their birth right when none of them want their sons to take up mountaineering? How did world acclaim and recognition impact the Sherpa's role in the growth of Indian mountaineering? And, what is the future of the community made famous by its climbing prowess?

In Darjeeling, the Sherpas are ordinary human beings with the same pre-occupations and individual strengths and weaknesses that make them, well, ordinary, just like the rest of us. But on the mountains, they are the magnificent men, there at the right time and with the right skills and attitudes that mark them as special from everyone else.

The Sherpa Project

Nandini and Deepa have been recording interviews and collecting stories of the climbing Sherpas of Darjeeling in order to put together a book on this unique community. The material is also being archived and digitally eternalized to preserve the information for Sherpas today and for future generations. Readers are invited to contribute any information or photographs they may have on the topic and may write to projectsherpa2012@gmail.com.

2

A Review of *the Himalayan Journal* Front Covers

W. M. (Bill) Aitken

Volumes 1 to 29 need not detain us because all have severe grey covers reflecting a sober age and the stiff upper lip values of British males. (Of the almost one hundred elected expatriate officers of the club, only Mrs H. P. V. Townend a vice president in 1939 breaks the mould of gender.) To prove that there was some daring locked up in those, for the most part establishment breasts, after the very early numbers of the Journal where the contents were printed on the cover imitating the fashion of all serious periodicals of that time, the style switched instead to print boldly on the cover full sized club logo depicting the chorten from Rongbuk. Magnifying a heathen symbol would have shocked the Anglican clergymen who in Halesowen conceived the Alpine Club some seventy years before the Himalayan Club saw the light of day in Shimla in 1928.

A chronological survey from the start of the era of illustrated covers to the present (volumes XXX to 68) reveals how the Journal's editors have all been willing to experiment and adapt to changing printing technology. Soli Mehta was the first of these pioneers and *HJ* XX introduced a half page black and white photo (of the Annapurna area?) without any caption perhaps because the modest editor had taken it himself. This and *HJ* XXXI both had a detachable jacket, creating the obvious problem that without the jacket the Journal was rendered nameless. The *HJ* XXX1 cover was a full page uncaptioned view of a high camp that heralded the nagging problem of where and how to overprint the Journal masthead for best effect without disrupting the photographic composition. In this case the sepia tinted photo allows clouds rolling in to swallow up the title. This issue marks the end of the

1. This is a personal look by a non-professional at the evolution of the *Himalayan Journal* front cover. It hopes to stimulate opinion that will result in enhancing the Journal's visual appeal. The illustrations, layout and typeface have been taken into consideration since only a successful fusion of all three elements makes for a memorable cover. (Ed)

2. *HJ* volumes were numbered in Roman till vol. XXXIV

THE
HIMALAYAN JOURNAL
Volume XXX, 1970

Volume XXXI
1971

THE
HIMALAYAN JOURNAL
Volume XXXII

THE
HIMALAYAN JOURNAL
Volume XXXIII

THE
HIMALAYAN JOURNAL

The
Himalayan
Journal

The
Himalayan
Journal
Volume 36

The
Himalayan
Journal

The
Himalayan
Journal

long relationship with the famous Baptist Mission Press of Calcutta: the printing job was transferred to Bombay (Mumbai) for editorial convenience. Following this transfer the use of Roman numerals would soon be dumped. Possibly someone pointed out that though the Romans had traded with the east coast of India, Mumbai would not appear on the map for another thousand years! Soli Mehta also abandoned the Journal's subtitle 'Records of the Himalayan Club'. Ominously the very next issue includes a page of corrections of which there are nearly as many factual errors as there are typographical.

HJ XXXII's cover marks the last half page black and white photo (featuring Baruntse.) The cover of *HJ* XXXIII has a close up of the summit of Machapuchare but the reproduction fails to do justice to the noble aspect. Or could the problem be traced to that perennial annoyance of scrimping on the office ink? The covers of volumes XXXIV to 46 are all of black and white photographs showing a steady improvement in quality by the increasing emphasis on the contrast between snow and rock. 'Camp 4 on Everest South West Face' the caption for *HJ* XXXIV well illustrates the precarious nature of its perch where a line of Whillans' Boxes seem to be straining at the pegs to glissade into the gloaming. For added effect there is a spectral setting sun, a favourite editorial aid that helped provide guts to their requisite glory. No less than six other covers (*HJ* 41, 45, 52, 53, 60, 68) include this visual Hallelujah effect that invokes heavens blessings to shine upon them. Having just survived three months of an unrelenting Mussoorie monsoon I appreciate the significance of the phrase 'a ray of hope'.

HJ 35 is the bumper golden jubilee issue with Harish Kapadia's colour photo of Nanda Devi from the south Sanctuary as cover. It introduces Harish as assistant editor and his influence is immediately apparent in the useful inclusion of a list of plates. This is an altogether highly pleasing cover where all the elements gell, despite the Journal masthead being placed in the middle of the Sanctuary meadows. Significantly the seriffed font is in lower case and does not in any way infringe on the grandeur of the two floating peaks of Nanda Devi. Later editions would miss this clue of how modest type size in lower case and an elegant seriffed font of subdued tone could make all the difference.

HJ 36 sees Harish emerge as full editor assisted by Roy Hawkins. The cover photo taken by James Wickwire along the northeast ridge of K2

The
Himalayan
Journal
Volume 39

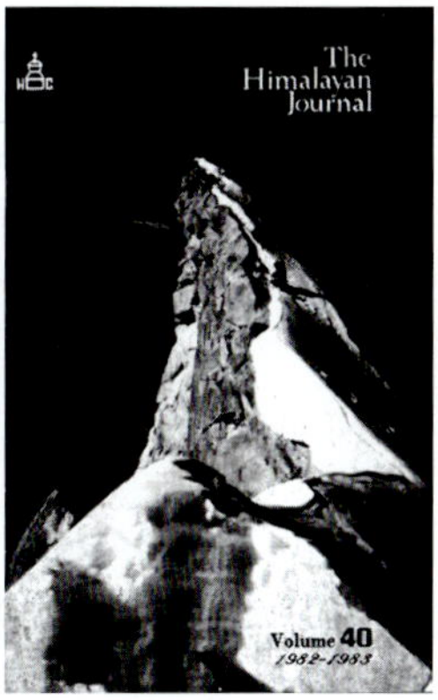
The
Himalayan
Journal
Volume 40
1982-1983

The
Himalayan
Journal
Volume 41

The
Himalayan
Journal

The
Himalayan
Journal
Volume 43
1985-1986

The
Himalayan
Journal
Volume
44
1986-1987

The
Himalayan
Journal
Volume
45

The
Himalayan
Journal
Volume
46

The
Himalayan
Journal
Volume
47

is a bit steamed up but at least succeeds in concealing the horrific voids all around. By contrast *HJ* 37 features on its cover the fluted column of Gauri Shankar captured by Major P. A. Cullinan, superbly reproduced. To compete in excellence of clarity is Rajul Mehta's superb black and white sketch of Sudarshan Parbat, a collector's item making it to the cover of *HJ* 38.

Bernard Odier's photographs often make it to cover display, a runner up being Chris Bonington. *HJ* 39's cover illustrates the tremendous fall of Gasherbrum II and on the front of *HJ* 40 Fotheringham's distinguished frame of Shivling's final sections showing Bonington reaching the summit provides a happy balance between climber and objective where the honours seem to be shared. The appearance of the *HJ* 41 cover shot of Tso Moriri lake by Romesh Bhattacharji is a bit spooky, more like a lowering Loch Ness than Ladakh. The famous photograph of the Loch Ness monster incidentally was taken by a gentleman emerging from a pub with a few drams inside him.

For cover No. 42 Harish has chosen Dhiren Toolsidas' fine study of the Rimo peaks in the East Karakoram. In soft contrast are the gentle summit slopes of Kumaun's Suli Top featured on the cover of *HJ* 43 in a photo taken by Ramakant Mahadik. The cover of *HJ* 44 – the diamond jubilee issue - presents a selection of Spiti peaks shot in striking contrast by Harish Kapadia, two on the front and four on the back cover. This issue marks an increase in the size of type for both masthead and logo and the beginning of a fatal flirtation with Madam Sans Serif.

Volume 45 cover has Chris Bonington's serene portrayal of the approach to Menlungtse which includes a corona effect. *HJ* 46 is something else and its cover photo of Thalay Sagar by Divyesh Muni makes a huge aesthetic impact. It is to my mind the pick of all the black and white covers. This issue records the passing of Soli and Hawk and the wisp of cloud above the mountain's summit seems to welcome the arrival of these kindred literary spirits.

The new era of colour photography from *HJ* 47 onwards comes as something of a mixed blessing. Colour undoubtedly has the capacity to provide authenticity but too often ends up as a pretty postcard scene. Colour has yet to achieve the wizardry of black and white in uniting opposites. So far no colour photographer has produced work whose

The
Himalayan
Journal
Volume
48
1990-1991

Volume
49
1991-1992
The Himalayan Journal

THE
HIMALAYAN
JOURNAL
Volume
50
1992-1993

THE
HIMALAYAN
JOURNAL
Volume
51
1995

THE
HIMALAYAN
JOURNAL
Volume
52
1996

The
Himalayan
Journal
Volume
53

The
Himalayan
Journal
Volume
54
1998

The
Himalayan
Journal
Volume
55
1999

The
Himalayan
Journal
Volume
56
2000

impact can compete with that of Vittorio Sella or Ansel Adams in capturing the living nature of mountains.

With colour came an increase in type size that extended to the club logo. Also the masthead accustomed to lower case in the black and white issues often appears in upper case. The cover of *HJ* 47, again by Bernard Odier shows the writhing Chogolungma glacier in the Western Karakoram to be a thing of living beauty and provides us with a timely reminder that the Himalaya is not just about verticality. In their constant reshuffling of scenes from various Himalayan regions taken at different altitudes, Journal editors through their choice of cover shots underscore the club philosophy of extending knowledge of Himalayan science, art, and literature as well as sport.

In several of the colour covers from *HJ* 47 to 68 I sense the larger and often un-seriffed type used for the masthead tends to compete with the photographic composition reducing the latter to background status and weakening the expectation that only a great cover can arouse. By printing the title in upper case on a snow field in red shadowed type, a dichotomy is created between subject and masthead. Modern taste in many fields seems to assume brutalism is the same as simplicity and stocky bland fonts are considered trendier than Baskerville or Sabon. To emphasise what is at risk of being sacrificed, the club published in 2001 *A Passage to Himalaya* edited by Harish Kapadia with a beautifully crafted cover, the photo of Aq Tash having been taken by the late Kaivan Mistry. The title is overprinted in Copperplate to give an exquisite final flourish. Kaivan's back cover photo (outlined in a fine yellow border) is also outstanding and for a back cover this is as tasteful as it can get.

Journal No. 48 waxes romantic in its cover shot of full moon on the Gasherbrum group but fumbles by inserting the title above the lunar orb. *HJ* 49 provides an original cover of the artist Serbjeet Singh's acrylic painting of a Himalayan panoramic overview. Unfortunately there is a mismatch between canvas and cover dimensions.

Wisely in my opinion the collage option has been foregone by the editors since the clarity of a single subject avoids the dog's dinner invited by resort to the manifold. The 50th jubilee cover does host a collage but the outlay in this instance is fairly restrained. Volumes 51 and 52 offer

The
Himalayan
Journal
Volume
57
2001

The
Himalayan
Journal
Volume
58

The
Himalayan
Journal
Volume
59
2003

60
THE
HIMALAYAN
JOURNAL

THE
HIMALAYAN
JOURNAL
Volume
61
2005

THE
HIMALAYAN
JOURNAL
Volume
62
2006

THE
HIMALAYAN
JOURNAL
Volume
63
2007

THE
HIMALAYAN
JOURNAL
Volume
64
2008

THE
HIMALAYAN
JOURNAL
Volume
65
2009

artistic variations on how to arrange a truce between title placement and photographic integrity. Bernard Odier achieves a cover hat trick with his sensitive rendering of the austere beauty of the Yarkand river but the bulky stencilled masthead mars the mood on the front cover (*HJ* 51). Snapping at his heels is Chris Bonington's groovy angle taken while climbing Drangnag Ri on the cover of 52. *HJ* 53 is the last experiment with a framed cover photo (till date) and presents Steve Berry's almost numinous shot of the chaddar walk on the freezing Zanskar river - in winter (as if we need to be told.)

Volume 54 has a nifty variant for a cover showing a digital route up Baruntse. The less said about the back cover the better. *HJ* 55 provides another eye catching cover shot by Harish of the Siachen glacier. So intense is the light radiated that it gives me the illusion that the range of mountains rising in the background has been morphed on to the glacier. Motup Chewang contributes the cover shot for *HJ* 56 viewed from a high ridge on Gya. The masthead placed in bold red amidships indicates how easy it is to fragment a cover's unity. To underline the proposition M. Taniguchi's cover photo (*HJ* 57) of Kanti Himal affirms how the presence of a single subject can be amplified by the subdued positioning of the title.

The worst victim of promiscuous placement is Mark Richey's potentially epic cover picture announcing *HJ* 58. Here the ultimate poetry of climbing gets rudely challenged by rufus interruptus. What deserves to be seen as a sensationally won insight on the grandeur of the Himalaya is reduced by the insertion of the masthead to become an Old Master's canvas unrecognised for its stand-alone genius. Sadly *HJ* 59 repeats the shortcoming and Hiroshi Sakai perhaps feels his cover photo of the Teram Shehr plateau has been similarly ruptured. No Japanese brush artist is ever guilty of such intrusion in their scroll paintings which are remarkable for the artistic inclusion of additional matter.

The 60th volume avoids disunity by the choice of yellow ink in Melvin Redekar's excellent close up of Cas van de Gevel against the glittering ice of Thalay Sagar, a satisfying cover complete with corona. Volume 61 is successful in diverting the masthead from the focus on the majestic passage of the river Siang taken on Harish's exploration of the Brahmaputra. *HJ* 62 makes another welcome change from the 'show us your muscles' content of many covers to frame the soothing feminine watercolour painting of Kangchenjunga by the sturdy medical missionary T. H. Somervell.

This interlude of reflectiveness is followed by two more intrusions of the red peril. Martin Moran's highly original capturing of Mount Kailash in Tibet along with its minor understudy in Kumaun (*HJ* 63) gets the disruptive treatment as does *HJ* 64. The latter cover depicts Mark Richey's quirky snow cave bivouac site on Latok I. To some it may convey the drama of high exposure but to me it conjures up the Mad Hatter's tea party in Disneyland. *HJ* 65 sees the abandonment of red ink for yellow and the result is an immediate reduction in visual confrontation. The cover shot of the stunning swirl of the Durung Drung glacier is by Harish presumably from the Pensi la. Extraordinarily this exalted view can now be seen from a bus window - though you may have to wait a week for the bus.

While Divyesh Muni's ethereal study of Thalay Sagar on the cover of *HJ* 46 is my black and white favourite, Martin Moran's close up of Rob Jarvis nearing the summit of Changuch (*HJ* 66) I find to be unbeatable as an expression of what human endeavour in the Himalaya is all about. This cover is the climax to years of editorial striving and makes the perfect farewell bow from Harish. You could argue the human presence in a photograph adds immediacy as well as scale.

The architecture of this cover is elemental where man (and title) have their appointed places and the above and below are at one. This is a classic symbol of the cocksure, an image that soars above the hostile opposites of Freud and the frigid. As it happens Martin Moran has also written my favourite Journal article in recounting his improbable trek to, and midnight climb of Nanda Kot after an immoderate feast, near the site where the Changuch photo was taken.

Rajesh Gadgil as Harish's successor, aided by the first lady associate editor Nandini Purandare, seem to be quick learners. Their two covers in my survey avoid the shortcomings and cash in on the experience so hardly won by Harish's long and devoted editorial innings. *HJ* 67's cover features Harish's shot of the elusive Kangto massif floating above a tribal stilted hut in East Kameng hinting at the mystique awaiting those privileged to tread this less known area of Arunachal Pradesh.

HJ 68's acrobatic cover of Renan Ozturk snapped on Garhwal's Shark's Fin against the diamond dazzle of the sun's rays is impeccably captured by Jimmy Chin. It echoes the lessons of layout learnt and crystallised in *HJ* 66. Together these two dramatic covers raise the bar and also the question whether the reader responds to the visually stunning or the inspirational.

As an amateur typesetter who purchased a tiny World War II letter press during the Emergency years these findings are open to question. Even if my reckoning is presumptuous it can serve to wrest from those better qualified opinions of more lasting worth. Reviewing all the covers makes you realise how nearly impossible it is for a single illustration to capture the elusive essence of the Himalayan experience: what makes us want to go there – and what compels us to go back. In spite of that a good many of the *HJ* covers succeed in pointing a finger in the right direction, surely the best (and overdue) tribute to the long hours of voluntary labour put in by the Journal's editors.

Summary

A personal view expressing impressions created by the front covers of the *Himalayan Journal* over the years by a well known author and regular contributor to the Journal.

3

Treks to Passes on India – Burma Border

Hoot Pass, Pangsu Pass and Chaukan Pass

Harish Kapadia

The Patkai range, in eastern India, divides India and Burma, with Pangsu pass as a low point on the ridge. On both sides of the range is thick jungle. The range is much smaller in height compared to the Himalayan range, but it is no less challenging in terms of travel. To the north, the range follows an easterly curve to meet the Kumon range joining it from Burma. This meeting of the ranges near the Chaukan pass has as much historical significance as the Pangsu. The Kumon range continues northwards to meet the Hkakabo Razi group of peaks near the Diphu pass. These two ranges divide Burma (in the east) and India (in the west). Apart from these major passes there are several other passes that allow passage across these ranges; namely Siddi, Hukawng, Shwangshan, Hapungan, Khumjawang and many other known only to shikaris like Hoot, Mugaphi, Gaphuka etc.

Frank Kingdon-Ward undertook many exploratory journeys on both sides of these ranges, crossing and re-crossing passes like Diphu several times. Those days borders between India, Burma and China were not strictly enforced and he happily roamed here in search of new mountains, botanical specimen and detailed scientific observations. We owe a lot to him for the knowledge related to these unknown valleys. He was trapped in the Lohit valley when, in 1950, a major earthquake (9.6 on the Richter scale) cut off these areas. For the first time a renowned scientist was present near the epicentre of an earthquake; who survived to record it for science.

Over the last two years I visited and trekked in the areas adjoining the border with Burma, reaching three passes.

Exploring Lapti Valley near Burma in 2011

The first approach to the Burmese border was through the Lapti valley in October-November 2011.

The trail is located in newly formed Anjaw District (on the Lohit river), Arunachal Pradesh. It is near the Rima-Kahao border with China on the Lohit river. Hawai is the new District Headquarter. If the sea and local dances grace the well known Hawaii (USA) in the Pacific, here at Hawai (Arunachal Pradesh), mountain scenery matches the beauty of the sea and the dances of traditional people are no less attractive.

This area contains the plateau opposite Walong, which is called the 'Burma Hump'. During the World War II several planes crashed on this plateau as old aircrafts could not gain necessary height and malfunctioned after a certain altitude. Remains of many planes lie scattered in the upper Lapti valley too, but now most of these remains have been scavenged by search parties and locals[1].

Ours was a beautiful exploratory trek. We descended steeply to the Lati river valley to reach the village Kamlat. After arranging porters we turned north from village Kamlat to its tributary, the Lapti. The trek was strenuous and the route climbed steeply- and as a result we had to descend steeply on the way back, on wet – slippery ground. But the forest and being on a remote trail to Burma was an inspiration. The immense beauty of the forest in autumn colours made camping grounds of Tafam and Kushok almost a paradise. The trail led us to Hoot pass which crosses into Burma and in few days would have lead us to Fort Hertz (Putao) in the Myanmar (Burma). F. Kingdon-Ward has written about the Lohit valley and the book *The Icy Mountains of Burma*, covers these mountains on the border. A few of these peaks, we observed on our way back. We came across two villages - they were almost a generation behind - and what we call 'progress' had not reached them as yet. No trekkers seemed to have come here though we heard some rumours of a party trekking here before. On the way back I fell almost 50 m (150 feet), on loose wet gravel covered by shrubs/bushes. I just could not hold on to the bushes, falling head first, gathered speed and passed over one slab after the other- all were downward sloping luckily. I then came to a halt on a small grassy patch and finally rolled over gently on a crop of huge rocks. It would have been a different ending if I had reached these rocks even at a slightly greater speed! Though I was bleeding profusely through my nose and was covered with many scratches, except for one sharp hit on my right hip I was saved. Dinesh Purandare reached me in a flash with few porters, and I could get up on my feet in about half an hour. I climbed up the steep slope to reach the

1. *HJ 65*, p. 62, 'Over the Hump' by Wing Cdr V. K. Sashindran.

main trail and then with the help of sturdy Mishmi porters walked down to the camp in a painful four hour trek. Medicines reduced the pain and next day I walked down to the roadhead from where a taxi took me to the Hawai rest house. A day of rest, two days of a rough car journey, a night on a train and a five hour flight followed before I was home!

These areas are wonderful, not visited by trekkers and have some of the finest virgin rain forest, leading to Alpine forests in the upper reaches. The high altitudes lakes are an attraction. It offers the most exhilarating experience and - not everyone has to have a fall.

Journey in 2013

In January 2013 I returned to the area with a different group of friends. We proceeded to the eastern-most point of India and along the India-Burma border. First was a ride on the oldest and eastern-most train route in India from Guwahati to Ledo, now a modern but deserted train station. Trains end here, but used to go up to Lekhapani a few kilometres ahead. Two kilometres ahead of Lekhapani, at 'Burma camp', began the famous 'Ledo Road' or 'Stillwell road'.[2] This road was built at great cost and labour across the Pangsu pass to Burma. It was named after Gen. Stilwell who was instrumental in its construction. The goods to support the war in Burma were loaded on trucks from here and were carried to Kunming, China - then an ally of the British. A lot has been written about this road so it was a nostalgic place to be and recall history.[3]

Deserted Lekhapani railway station. Most returning soldiers from Burma during the war finished their horrifying journey here and caught train to safety. (Lakhshmi R.)

2. There is confusion in names. Lekhapani is the name of the last railway station, starting point of the Ledo road and home of army camp from where the Burma war effort was helped. Likabali is an army station, north of Brahmaputra, across from Dibrugarh. Both are in Assam.

3. *The Burma Road,* by Donovan Webster. (Pan Macmillan, Oxford, 2003)

Second World War cemetery at Digboi. (Harish Kapadia)

From Ledo, at first we drove back to Margherita, also with nostalgic mentions in history. Now it is a sleepy little town which has no charm except its catchy name. Ahead was Digboi, where the first oil well in India was discovered. As the story goes, when the well was being dug a British officer riding a horse encouraged the workers by shouting ‘Dig boy, dig’. This was changed to Digboi, the present name. Oil was struck on 19 October 1889 and the well was abandoned in 1932. An oil museum and a well kept war cemetery nearby are worth a look.

The first oil well in India at Digboi. It functioned from 1890 to 1927. (Harish Kapadia)

We spent a night at Jairampur (named after Jairamdas Daulatram, former benevolent Governor of Assam, which included Arunachal Pradesh then). Nearby was a cemetery recently discovered in the forest, containing many graves of the allied soldiers. It had remained hidden in thick foliage. After another night at Nampong, we drove to the Pangsu pass, on 10 January 2013 to cross into Burma.

Pangsu Pass (1136 m)

During the Second World War, the Japanese advanced into Burma in 1942. They overran the allied forces, forcing them to retreat in a hurry. It was a very disorderly retreat in face of the fast advancing enemy.

To reach the safety of British India, refugees and the retreating army had to cross the Pangsu pass, on the Patkai range, which divides India and Burma. It was a hard and disastrous journey. Soldiers and several civilians had to trudge through deep sticky mud, a curse of this frontier. Many times they sunk till their waists in the soft mud, going on without much food, ravaged by malaria, lashed by the monsoon and a variety of ailments. The rain forest was dense and the rivers were flooded. A large number of refugees came across the Pangsu and though it was a lower pass, it was a steep climb through knee-deep mud. This historic gateway was nick named the 'Hell gate' by soldiers.

Pangsu pass at India-Burma border. After long arduous trek the British soldier escaped from the pursuing Japanese army though this pass. This board was the site of their cry of freedom. (Harish Kapadia).

The Pangsung village was about three km inside Burma.[4] The road was atrocious and one is not allowed to drive on the Burmese side. Burmese boys were ferrying goods from the Pangsu pass to the village on fast moving motorbikes. We enjoyed Burmese food and saw the small bazaar that had sprung up for this day. Burmese military boys (yes, they were very young) checked our permits and ensured that no camera or mobile phone was carried inside their territory. Now the road is being widened to a four lane highway and is intended to

4. Indians are allowed till Pangsung on 10, 20 and 30 of every month. Burmese are allowed into India till Nampong on every Friday.

The famous 'Lake of No Return' in the Burma plains, seen from Pangsung village in Burma. (Dr Nanak Bhagat).

be the Asian Highway connecting India with southeast Asia. Looking at its state today this is a far away dream.

Lake of No Return (Nawng Yang in Burmese)

It was nostalgic to see the 'Lake of No Return', in the plains of Burma from here. It was a vast lake, the only clearing in thick forest. During monsoon surrounding areas would fill up with water too, making it a giant lake. When they developed engine trouble, the planes flying across the 'Burma Hump' tried to land on these waters, their best chance of survival amidst thick forest all around. Most of them did not return to tell the story.

Other stories also explain the name of 'Lake of No Return'. The second has it that a group of Japanese soldiers returning from battle lost their way and ended up at the lake. There, they were stricken by malaria and died and hence the name. According to a third story, US army soldiers, working on the Ledo road, were sent to examine the lake, got trapped by the undergrowth and perished. A fourth story says this is the 'lake of no return', because retreating British troops in 1942 got lost in quicksand. Adding myth to legend, one author claimed he had encountered the name on a document written by one of the Ten Lost Tribes of Israel, which he claimed were still hiding out in the area.[5]

By late afternoon we returned to Nampong and drove a short distance to Miao (via Jagun) to prepare for our venture to another historic pass, the Chaukan pass. Miao is a major administrative centre with government and Namdafa National Park offices. Permits had to be obtained from here. It has rather mixed population as Chakma refugees (Bangladesh refugees) had settled around Miao since 1971.

The Epic of Chaukan Pass

> Wild elephants are known to have used Chaukan Pass as an ancient migration route linking populations in Assam and Burma. The giant footprints of the pachyderm are still more likely to be spotted in stream side mud than those of homo-sapiens. This is

5. An Indian newspaper *The Telegraph* reported, in a story on the possible reopening of the Ledo road in 2007, that 'close by [Pangsu] is the Lake of No Return — the local Bermuda Triangle. According to folklore, aircraft that fly over the lake never return.' The lake's reputation is advertised in hopes of making the area more attractive to tourists? Who knows, the 'Indian' Bermuda Triangle might just turn out to be the next tourist-puller of the region!

one of the areas of the world in the twenty-first century, which can claim to be largely unexplored. As one of those who survived the desperate journey of 1942 wrote afterwards:

> 'Now we know why this stretch of country is uninhabited for 100 miles. Not only there is no trace of men, but mammal and even bird life is conspicuous by its absence: truly a forgotten world, where solitude reigns supreme.'
>
> Diary of Millar IOR/M/3/1180
> *Exodus Burma* by Felicity Goodall, p. 214

One of the groups fleeing the Japanese army consisted of the British who were trapped at Fort Hertz (now known as Putao) in north Burma. Advancing Japanese troops may have overrun Fort Hertz if they had waited till the monsoon period passed. So in heavy rains they left Putao with a Mishmi guide and porters to cross the Chaukan pass, nearest to their location.

They had to come across the junction of the Patkai and Kumon range through dense forest in heavy rains. More than 230 British people crossed over from here with Sir John Rowland including a three month old child. They had to flee in June, when it was warm but at the height of monsoon, with flooded river, leeches, snakes and malaria mosquitoes. Once across the pass their progress was almost reduced to a crawl, covering just about two / three km per day. There were no villages and the trail almost unknown. Finally they were stopped by Tilung Hka, the first major tributary of Noa Dihing.

A party of two persons had gone ahead and reached Miao, to alert the authorities about the plight of the trapped party, and a rescue was organised. Food was air dropped to them whenever the weather was clear. A party led by Gyles Mackrell, from the Indian Tea Association of Assam, mounted a major ground rescue with several elephants. He was accompanied by troops from Assam Rifles, all sturdy Gorkhas who excelled in helping the rescue efforts. At one point the refugees were stranded on an island near the Dapha Hka (river). Elephants with the Mackrell party reached the island and rescued all of them. The island was soon washed away. This rescue saved many lives. As they neared Miao, ground support increased with food sent with porters and progress to safety was faster.

Army team to Chaukan Pass, 1961. Route is more difficult in 2013, due to floods and erosion.

The first batch on the Indian Army which visited the area and reached Chaukan pass in 1961. (From Kunal Verma collection)

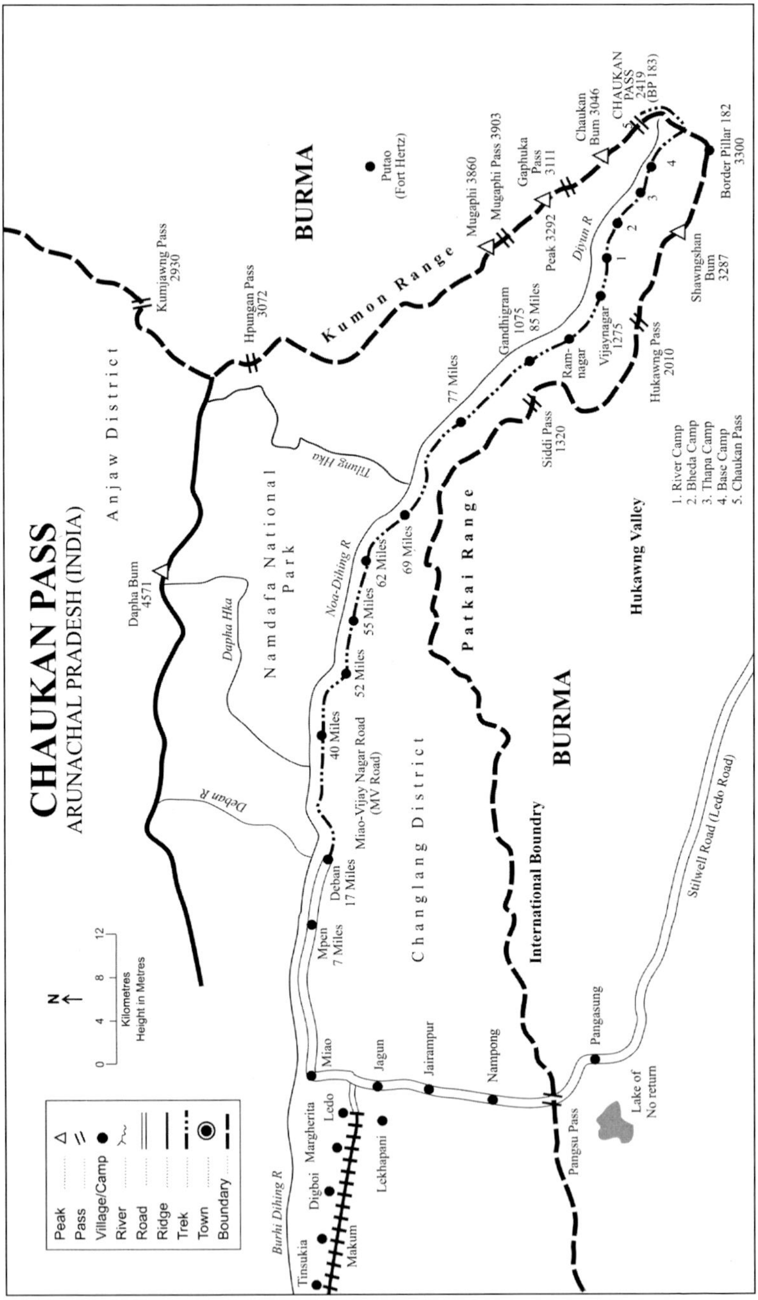
CHAUKAN PASS
ARUNACHAL PRADESH (INDIA)
Peak
Pass
Village/Camp
River
Road
Ridge
Trek
Town
Boundary
N
0 4 8 12
Kilometres
Height in Metres
Anjaw District
Namdafa National Park
Changlang District
Patkai Range
Kumon Range
BURMA
BURMA
Hukawng Valley
International Boundry
Kumjawng Pass 2930
Hpungan Pass 3072
Dapha Bum 4571
Dapha Hka
Deban R
Tiding Hka
Noa-Dihing R
Diyun R
Burhi Dihing R
Putao (Fort Hertz)
Mugaphi 3860
Mugaphi Pass 3903
Gaphuka Pass 3111
Peak 3292
Chaukan Bum 3046
CHAUKAN PASS 2419 (BP 183)
Border Pillar 182 3300
Shawngshan Bum 3287
Hukawng Pass 2010
Siddi Pass 1320
Gandhigram 1075 85 Miles
Vijaynagar 1275
Ram-nagar
77 Miles
69 Miles
62 Miles
55 Miles
52 Miles
40 Miles
Miao-Vijay Nagar Road (MV Road)
Deban 17 Miles
Mpen 7 Miles
Miao
Jagun
Jairampur
Nampong
Pangasung
Pangsu Pass
Lake of No return
Stilwell Road (Ledo Road)
Tinsukia
Makum
Digboi
Margherita
Ledo
Lekhapani
1. River Camp
2. Bheda Camp
3. Thapa Camp
4. Base Camp
5. Chaukan Pass

These efforts by men and animals saved many though about two dozen perished. Finally a haggard party reached the Lekhapani-Ledo railway line and onwards to safety. This was the Epic of Chaukan pass.[6]

In 1961, a year before the Chinese invasion, the Indian army mounted 'Operation Srijitga' to reach Chaukan pass. They too encountered many difficulties but the party led by Maj. Gen. Guraiya finally reached Chaukan pass on 27 November 1961 and planted the Indian flag.[7] After discussions with the Burmese authorities, Border Pillar No. 183 was erected here in 1971 with the 'Watershed Principle' being followed. All rivers from this horse shoe basin flow west-northwest into the Noa Dihing, which then flows into Assam as the 'Buri Dihing'. It then merges with the Lohit river and finally into the Brahmaputra.

To occupy the area permanently many Nepali families of retired Assam Rifles personnel were settled here with full government support. Large grants and land was given to them. Food was supplied to them through helicopters and AN 32 planes, (even today) and the Miao-Vijaynagar road was built in about 1972 (not fully functional till today). Several native Lisu tribal families, who were originally migrants from Burma, also settled here over decades, much before India-Burma boundary demarcation.

During February to April 2000, a party of scientists from Anne Mane Foundation, Bangalore wanted to research routes of elephant migration between India and Burma. One of their aims was to reach Chaukan pass. Due to difficulties they stopped at least three days before reaching the pass.[8]

Our Epic of Chaukan Pass

We drove to the Deban tourist lodge located in the Namdafa Sanctuary. Facing Namdafa river and on the banks of Noa Dihing it was well situated to observe birds and wild life. Dapha Bum peak (4570 m) is seen towering over the river valley from here.

6. *Forgotten Frontier,* by Geoffrey Tyson. (Spectrum Publications, Guwahati, 1992) and *Flight By Elephant,* by Andrew Martin.(Fourth Estate, London, 2013). See the movie 'The Elephant Man' (rescue of Chaukan pass refugees, Cambridge University), on YouTube.

7. *The Children of the Dawn. Northeast Triology,* by Dipti Bhalla and Kunal Verma. (Kaleido India, Gurgaon, 2011).

8. See the film 'The Old Elephant Route' by Anne Mane Foundation.(YouTube)

The Noa Dihing river from air. (Kunal Verma)

The Noa Dihing was a beautiful river, now serene and blue, surrounded by forest. In monsoon it must be a terror with flooding water and crumbling sides. The valley is adjoined on three sides by the Burmese territories and for generations it was ruled by the Ahom kings of Assam. This land was occupied by the Singhpos, who were akin to Kachins of the Hukawng valley of Burma.

In May 1825 Burmese forces advanced through the pass to re-conquer it from Assam. The Singhpos who dealt in slaves, a policy not tolerated by the British, feared that they may be expelled from land they owned in the Brahmaputra valley, so they turned against the British and joined the Burmese forces. The invaders had reached lower Noa Dihing and built defences there. They were met by Captain Neufville with 300 Sepoys and two gunboats at Dapha. A strong stockade position was held by many Burmese soldiers and few cavalry. Capturing these positions the British force proceeded upstream. They met a larger force of Burmese and Singhpos at Bisa, about two miles upstream on a tributary of Noa Dihing. In a brutal fight Burmese were defeated and Singhpos scattered. For next decade the British held all the passes on the Patkai range and cleared the valley from occupation by the Singhpos and Burmese.

Later Daphna Game, one of the four prominent Singhpo chief crossed the Patkai range from Hukawng valley and attacked Bisa. The British reacted strongly and repelled the rebel after a stiff fight and losses on both sides. A permanent British post was now created at Bisa. Most of the Singhpos returned to Hukawng valley in Burma where slavery was prevalent.

A detailed survey for an India-Burma railway line was undertaken in 1896. A strong force with the surveyors crossed the Patkai range to the Hukawng valley and joined it with a similar survey party in upper Burma. However the survey for a railway line remained on paper only.[9] The Noa Dihing valley remained part of the British Empire and later became part of India.

On 13 January 2012 we drove on the Miao-Vijaynagar road or 'MV Road' for a short distance. We first stayed at 40th mile; all stops are named as such. We met our Lisu porters here and now on we had to walk about 117 km to Vijaynagar. The first taste of this road was enough to send the weak-hearted back home. It was a hell of a road, with deep sticky mud everywhere. Powerful four wheel drive trucks had to be pushed by bulldozers to proceed. While walking if you get off the road you can slip in a culvert and if you step on the other side it was in knee-deep mud. Shoes were stuck in, walking poles fell apart and clothes were as dirty as they could ever be.

Walking on muddy Miao-Vijaynagar road. (Harish Kapadia)

On the second day we went down to the river and crossed the 'Burma nala' on a swinging bridge. The route ahead, along the river, was a delight to walk for few km, passing the Lisu village of Ngawazah where we tasted the local fruit Thazi (Persimmon

9. *History of Upper Assam, Upper Burmah and North-eastern Frontier,* by L.W. Shakspear. (Spectrum publications, Guwahati, 2004).

The 'Burma Bridge' across the 'Burma nala' now and in past. (Lakshmi R.)

or Sharon). Crossing the Noa Dihing on a couple of scary stilt bridges, called dahlons, we reached the village Nibodi ('52 miles'), where we camped on an open ground. Next day we had to return to the muddy MV road. For the next three days we were on the road, camping on clearings (at approximately 55 miles, 62 miles and 69 miles) by the side of the road with a water source nearby. Small thatched huts of locals along the road sold tea, biscuits and some sundry stuff. Large fish were seen hanging in many of these shops available for sale. After 62 miles, a futile attempt to hitch a lift on a truck ended in disaster when the truck itself got stuck and even the bulldozer could not move it. On the fifth day of the trek, we descended from the road to a beautiful campsite right on the river bed close to the village Sichodi through which we walked past the next day. It is a wonder how these remote villages survived with meagre supplies and in heavy rains with such a muddy road.[10]

From Sichodi, climbing steeply through the jungle, we were back on the road and back to mud. The valley opened up as we neared Gandhigram, named after a small statute of Mahatma Gandhi installed here. It seems decades ago two or three statues of various leaders arrived for installation. No one could recognise the statue of Gandhi from the lot.

10. These villages are part of Namdafa National Park. This is their traditional home and any suggestion to shift from the area is strongly rejected by them.

The small statue of Mahatma Gandhi at village Shidi. This village is now named as Gandhigram. (Harish Kapadia)

Finally another statue was flown in with words 'Gandhiji' written on the back, which can still be seen. The original name of the village 'Shibudi or Shidi', was changed to Gandhigram. We spent a day at Gandhigram going around the well laid out village and eating some delicious locally grown pineapple and sugarcane.

It was a peaceful village and with friendly Lisu villagers. Ahead there were the villages of 'settlers'- Nepalis retired from the Assam Rifles, with names like Ramnagar. They looked prosperous, a pleasant surprise, as the only contact with civilisation was through irregular helicopter flights. Above the main road seven km to the northeast is Pritnagar, a small village known for its beauty with a small lake.

The last 18 km road travel to Vijaynagar was relatively pleasant. We spent a night at Vijaynagar, at the Assam Rifles post near the airstrip. The original name of this village was Dawodi but the name was

Crossing traditional 'Dahlon Bridge' across the Noa Dihing river. (R. Wani)

Panorama of Vijaynagar. (Dr. Nanak Bhagat)

changed in honour of Maj. Gen. Guraiya's son Vijay. On northeast was the Kumon range and southwest the Patkai range, and almost where both these ranges meet was the Shawngshan pass (and peak) (3287 m). This pass southeast from Vijaynagar, led to the Hukawng valley of Burma while the Chaukan pass led to Putao in the northeast. It was said that relics of a Buddhist stupa and images had been discovered in the vicinity of Viyaynagar. No one had any knowledge of it now. The area was earlier known as Jahu-Natu.[11]

Final Approach to Chaukan Pass

That night at Vijaynagar we sorted our gear and replenished some of our essential rations. One young guide from Assam Rifles was hired. He knew the route but with youthful energy he always went ahead at 'army speed' leaving us looking for him!

Jungle and a Woman

> .....jungles vary so considerably, that it is difficult to give an apt general description. Perhaps the best way is to compare the jungle with a beautiful woman, the pin-up dream girl in full technicolour, cool, alluring, attractive to look at, but, once approached and negotiated with, full of greatest possibilities of danger and death to the unwary. This simile is particularly apt because, like the figure of the pin-up girl, the jungle is never flat.
>
> Lieutenant Colonel O.G.W. White DSO

---

11. *Gazetteer of India, Arunachal Pradesh, Tirap District*. Ed. S. Dutta Choudhury, Govt. of Arunachal Pradesh, 1980.

True to the above description, the route ahead was beautiful, exploratory and dangerous. This was the real stuff- the trek on a thin trail along the Diyun river (as the Noa Dihing river is known here). Nothing had changed in last seven decades in terms of trail and difficulties en route. Climbing steep ridges, descending wooden ladders, crossing single log bridges required a good balance and a strong heart. We constructed about 16 small bridges to cross the river to avoid steep climbs on either side. Our Lisu porters, especially Gisa as the chief engineer, were marvellous at this local engineering. It was tiring and tense all along as we had to climb and descend regularly. At many places the trail had to be cut, log bridges installed at sides to cross along the bank and cross many fallen tree trunks. We stayed at river camps, Bheda, Thapa and finally at the base camp, all names given by the Assam Rifles patrols, the only ones to visit the pass annually. Being January it was cold at night but with so much wood around and supported by strong, helpful and energetic Lisu porters nothing can stop you. The forest was both a joy and fear. Not many animals or birds were seen or heard – possibly the Lisus had eaten them all for generations! However we could hear calls by Gibbon monkeys, saw few hornbills and there was much fish in the river. The advantage of trekking in winter was evident: no leeches, no snakes, no dim dim flies and low water level. Finally we were at the foot of the pass, which the army had called 'base camp'. From here the river continued southeast while our trail climbed steeply along the adjoining ridge. Once at height the trail followed the ridge to descend across two false passes to the pass.

Border pillars at Chaukan pass. (Harish Kapadia)

On 27 January 2013 we left the base camp early and climbed almost 1000 m to the top of a ridge. Then we descended along a forested ridge for more than 500 m and traversed through Burma in the east. Finally after a 10 hour day we further descended to the historic Chaukan pass (2419 m) and camped near the Border Pillar no. 183 which marks the border between India and Burma, some tents in India and some in Burma, but just a few feet apart from each other!

After almost 70 years since the crossing by British refugees and 52 years after 'Operation Srijitga', when a patrol of Indian army reached the pass, we were the next (civil) party reaching the pass. When we started little did we know that the difficulties had multiplied manifold due to floods, landslides and erosion over these decades.

It is said that the pass was a route that wild elephants used to cross. The pass, where it is marked today, appeared too steep on either side for such a passage. Little before the Border Pillar a wider and gentler crossing point was seen. This appeared to be feasible for such a crossing. Moreover the Diyun river originated near this gentler pass. But only an expert in elephant migration can make accurate judgement. On 28 January we reversed the trail from the Chaukan pass, climbed up 500 m and descended 1000 m. In all we had an ascent and descent of almost 3100 m over two days to fulfil our dreams.

Systematically and with much concentration we carefully covered the return trail over our newly made bridges, muddy forest trails and slippery logs. Finally we were at Vijaynagar. After a two-day wait we were able to catch the helicopter sortie which supplies material here. It was a 60 minute flight to Mohanbari /Dibrugarh, which otherwise on foot, would have taken us at least a week to cover. We flew over many historical places like Hoot pass (to north), Chaukan pass (to southeast), Putao (in far northeast), and Pangsu pass (to south). As the helicopter landed at the modern airport, we stepped out of history into the present.

Summary

Treks to passes on the India-Burma border. Hoot pass (3570 m), Pangsu pass (1136 m) and Chaukan pass (2419 m).

Members: 2011: Dinesh, Nandini and Uttara Purandare, Atul Rawal and Harish Kapadia.

Period: 25 October to 06 November 2011

Members: 2013: Rajendra Wani, Lakshmi R., Dr. Nanak Bhagat and Harish Kapadia.

Period: 07 January to 02 February 2013

4

Trisul - the West Ridge

Brigadier Ashok Abbey

On 27 September 2012, a team of soldier mountaineers commenced moving up the Nandakini valley from Sitel, the roadhead in the Chamoli district of Uttarkhand. The team was late, in fact too late to be moving up at this time for a post monsoon climbing expedition. Deep frosts on the Bugyals, low temperatures, shorter days in the valley were some ominous signs of an impending early winter. Their destination, Trisul I via the west ridge seemed far and distant, lost in the perpetual low cloud base, which engulfed the horizon. The team established base camp at Hom Kund on 2 October, only to be greeted by the first heavy snowfall of the season on 4 October. Nevertheless, the team forced its way up the lower ramparts of the mountain, through a treacherous couloir and established Camp 1. As the expedition stepped on the Raunthi Bank after Camp 1, they knew that it was time to turn back, but not before they carried out a detailed reconnaissance of the upper Raunthi Bank and the majestic west face of Trisul I, towering above. It was then that the team decided to get back for another attempt on Trisul I for a pre-

Trisul I west face, as seen from Camp 1and Raunthi Bank. (Brigadier Ashok Abbey)

monsoon expedition in May-June 2013, albeit with the objective of making the first Indian traverse of the three great peaks of Trisul.

The Trisul massif stands majestically in the Chamoli district of the young Himalayan state of Uttarakhand, as one of the most fascinating mountains of the Garhwal Himalaya. Mired in mythology and deeply revered, the mountain lies in Patti Talla Panikhanda, Pargana Panikhanda and is located on the outer wall to the south of the Nanda Devi Sanctuary. The British Garhwal Gazetteer interestingly, describes the massif as follows:-

> Trisul is the trident of Siva. The three peaks, which form the trident, lie almost in a straight line running from north-east to south-west. The highest of all is situated at the north-eastern extremity of the ridge and has an elevation of 23,406 feet. The summit is described as a flat-topped dome of snow forming the apex of a huge triangular snow field set at steep angle on the north-east face of the mountain. Between this peak and the middle peak (22,490 feet) lies the Trisul glacier and beyond the latter again lies the third peak, 22,360 feet.

The Shivpurans describe Trisul as 'Shiva's might' or a 'symbol of power' used by Lord Shiva, as the ultimate weapon for destruction of evil. It is also regarded as a divine instrument, used to regulate the forces of life and to maintain balance and harmony of all the three worlds. The mountain makes fascinating viewing as it takes different shapes and form, when viewed from Bedini Bugyal, Ranikhet, Kausani, Nandakini valley and other places in Garhwal and Kumaun.

Trisul I standing at a majestic altitude of 7120 m, is the highest and the most dominating peak of the Trisul massif. The geography of Trisul massif is complicated with a series of ridges emanating in virtually all directions. The northeast ridge from Trisul I continues unabated for nearly 10 km culminating at Bethartoli Himal North (6352 m) via Bethartoli South (6318 m). Five km north of Trisul I, from an Unnamed Junction Peak, the ridge swings northwest towards Bethartoli North and South, and finally merges with the valley floor of the Rishi ganga towards the north. After approximately five km from the Unnamed Junction Peak, from where the ridge bifurcates towards Bethartoli South and North, the second half of the ridge continues unabated in a north easterly manner hemmed between the Bethartoli Bank to the north and the vast Trisul Bank to the south. It overlooks the Nanda Devi Sanctuary towards Tridang and the Devasthan[1] massif comprising of Devasthan I (6490 m) and II (6465 m). The east ridge of the mountain looses height over a distance of three km and towers above the Trisul Bank.

The southwest ridge of Trisul I continues for nearly over a kilometre, before turning westwards and taking form of the west ridge. The west ridge of Trisul I after breaking from the southwest ridge subsequently turns west and then northwest and cuts across the upper Nandakini valley, overlooking the Hom Kund and the Sili Samudra glacier valleys, to the east and the south. The southwest ridge, after emanating from the peak, breaks off from the west ridge after one and half km and cuts across southeastwards, rising to the summit of Trisul II. The distance between Trisul I to II, as the crow flies is approximately two km. From Trisul II, the south ridge comes down over a complicated broken ridge line of nearly three and half km, riddled with objective dangers before dropping to a col and then rising to the summit of Trisul III. The Trisul massif thus stands as the great divide between the waters of Rishi ganga, Nandakini and the Kali river valleys.

Trisul I (7120 m) was first climbed in 1907 by Dr Tom Longstaff along with his Gurkha assistant, Subedar Karbir Burathok of the 5th Gorkha Rifles and two Swiss guides, the Brocherel brothers. This was then an altitude record and the highest summit to be climbed in the world. In 1951, Roy Greenwood, a physical training instructor at the Indian Military Academy, Dehradun along with Gurdial Singh (Guru), climbed[2] the mountain from a westerly direction, towards the summit from the north ridge. The headstand on the summit of Trisul I by Guru became legendary, more so as this was the first major climb by an Indian of Independent India, thus heralding the beginning of Indian mountaineering. Interestingly, this is what the then editor of the *HJ*[3] commented post the climb:-

---

1. The Survey of India records the name of this massif as Devasthan and not Devisthan as mentioned elsewhere.
2. *HJ* Vol. XVII, p. 110
3. *HJ* Vol. XVII, p. 112

> Roy Greenwood and Gurdial Singh were at the time on the staff of the Indian Military Academy at Dehra Dun. This was the first time that an Indian member of the Himalayan club had ascended one of the greater Himalayan peaks, and it is to be regretted that we are not able to record his own observations and reactions; but we take this opportunity of congratulating him on a very gallant effort.

In 1960, a Yugoslavian expedition (A. Kunaver)[4] made the first ascent of Trisul II (6690 m) and III (6609 m), after approaching the massif from the Bidalgwar glacier. In 1978, a Japanese team (Sadashige Inada)[5] climbed the mountain from the Trisul Bank, over a twenty-eight-day period. They made an ascent of Trisul I from the south ridge and from the north ridge to Trisul II. In 1987, a strong Yugoslavian team (V. Lado) returned and climbed Trisul I from the west face. They also climbed Trisul II and III from north to south, thus recording the first and the only traverse of the great peaks of Trisul till date. Ever since, a number of expeditions have climbed Trisul I, majority being from the longer but easier traditional approach from the north (now in the Nanda Devi Sanctuary), including a ski descent. In the recent times, a few expeditions have attempted and reportedly climbed the mountain from the challenging west face.

The 16 member Indian Army team from the Central Command, sponsored by the Army Adventure Wing, which attempted a traverse of the Trisul massif in Garhwal Himalaya in 2013, was a blend of experience and youth. Majority of the climbing team members were former instructors and trainees of the High Altitude Warfare School, Gulmarg. The aim of the expedition was to make an ascent of Trisul I (7120 m) from the west and southwest ridge and then traverse south to Trisul II (6690 m) and onwards to Trisul III (6008 m). It is pertinent to mention that porters were only used for stocking up to base camp. No High Altitude Supporters or Sherpas were used by the expedition beyond base camp.

The expedition after congregating at Joshimath on 09 May moved along the Alaknanda river till Nandprayag. From Nandprayag, which

4. *HJ* Vol. XXII, p. 70

5. This yatra (pilgrimage) is held every twelve years.

Looking from the upper west ridge. L to R - Trisul II, III , Unnamed peak and Chandini Kot. (Brigadier Ashok Abbey)

is the confluence of Nandakini and Alaknanda rivers, the team moved along the Nandakini river via Ghat, Sitel, Sutol, Tatra, Lata Kopri, Bhujani, Chandnighat and Dudhang. Leaving the Sili Samudra glacier to the southeast, the expedition moved along the upper Nandakini river to Hom Kund, which was the base camp. Subsequently they gained the upper Raunthi Bank to establish Camp 1 and 2 and ascended the west ridge to establish Camp 3, en route to the summit.

Base camp was established on 18 May, after leaving the roadhead camp at Sitel on 11 May. The site overlooked the spectacular, high altitude lake of Hom Kund, which is a deeply revered site and the culmination point of the Great Raj Jat Yatra . Base camp was located at 4110 m, three km south of the Raunthi saddle, under the shadow of a mammoth icefall emanating from the upper Raunthi Bank to the east.

From base camp, the team ascended a couloir to the east. This was a dangerous steep funnel, which fired a volley of shooting stones regularly from its upper ramparts. The couloir, which at the time of induction was covered with snow, soon revealed its loose belly that of a very dangerous scree slope. Three ropes were fixed in the upper parts of the couloir to make it safe for movement. Moving with caution, Camp 1 was established on 23 May, at 5120 m. The camp overlooked the upper Raunthi Bank, under the inspiring west face of Trisul I, with the Bethartoli Himal massif rising to its north.

Camp 2 was established and occupied on 25 June 2013, at nearly 6000 m. The route from Camp 1 to Camp 2, was along the western extremity of the upper Raunthi Bank. Initially, the team moved south close to the ridgeline overlooking the Hom Kund and the upper Nandakini valley. The route then veered southeast wards. After crossing the heavily crevassed ridden snowfield with caution and keeping well away from the west face, the team established Camp 2 in the upper icefield of the Raunthi Bank. Camp 2 was a dangerous location, prone to ice avalanches. On the night of 31 June, one of the ice walls of the west ridge collapsed at 01.00 a.m. The dangerous ice splinters hit three tents at the camp, ripping two of them. The members had a narrow escape and were fortunately unscathed.

The west ridge of Trisul I merges into the upper Raunthi Bank. From Camp 2, the team moved up to gain the last snow field of the glacier. From here, the formidable west ridge which separates the Raunthi Bank from the Sili Samudra glacier nearly 2000 m below, takes shape and form and is clearly visible. The team under Naib Subedar Madan Singh negotiated a series of ice walls (60^0 and above) including patches of blue ice, to gain the crest of the west ridge. Camp 3 was established on 28 May on the spine of the west ridge. Two tents were pitched after cutting ice platforms, which were secured and anchored firmly. The mountain plunges into an abyss towards the south.

After establishing the camp, the team carried out a reconnaissance of the west ridge and worked on the route above Camp 3. Camp 3 at 6420 m was finally occupied on 29 June 2013, by the first summit team. Approximately 700 m of rope was fixed, between Camp 2 and Camp 3. On 30 May, a team comprising of Naib Subedar Madan Singh, Hav Matwar Singh, Lance Naik Min Bahadur Tamang, Rifleman Yashpal and Paratrooper Ashok Kumar left Camp 3 at 02.30 a.m. This first attempt was supported by Brigadier Ashok Abbey, Subedar Topgey Bhutia, Assistant Leader Kunchok Tempa, Assistant Leader Tenzin Rigden, Naik Bishnu Kumar and Rifleman Birendra Singh at Camp 2, who moved up the same day.

The west ridge of Trisul I is a long, sharp, rising, broken ridge, lined with unstable ice walls. To the north and northwest, the mountain drops towards the Raunthi and Upper Nanda Ghunti Bank. From Camp 3, the team climbed the exposed west ridge with caution. Two ropes were fixed from Camp 3 upwards, as the team negotiated a crevasse and climbed the ice wall near the first prominent rock. It took them nearly 90 minutes to reach this landmark. From here, the team took nearly three hours to the second prominent rock. Just short of the point, where the northwest ridge from Trisul II meets the west ridge of Trisul I, the team decided to rope up. It took over an hour from the rock face to the end of fixed rope. After roping up, they moved for another 70 minutes to reach a small snowfield, at the base of the final summit pyramid. The summit of Trisul I was heavily corniced at this time, with the lip looming menacingly towards the east and southeast. The cornice was dangerous and the team approached the final summit, climbing the summit pyramid from the southwest with great caution. At 09.20 a.m., they stopped short of the summit due to the dangerous cornice and the unstable snow conditions. They subsequently retreated to the base of the summit pyramid. Enroute, despite the diminishing visibility and cloud build up towards the late morning hours, they had good views of Nanda Devi and Nanda Devi East, towards the west. Splendid views of the ridge connecting Trisul II and III as also the peaks of Mrigthuni (6855 m), P. 6310 m and others to the southeast were visible. The team finally descended to Camp 2 and onwards to Camp 1, reaching the site by 06.00 p.m.

On 31 June 2013, a team comprising of Brigadier Ashok Abbey, Subedar Topgay Bhutia, Assistant Leader Kunchok Tempa, Assistant Leader Tenzin Rigden, Naik Bishnu Bahadur and Rilfeman Birendra Singh occupied Camp 2, for a direct attempt on Trisul II. The team reached Camp 3 after a gruelling ascent in poor visibility at 03.30 p.m. on 30 June. Towards evening, low clouds engulfed the entire valley.

As per the original plan, the team would move on the fixed ropes up to junction of the northwest ridge emanating from Trisul II, with the west ridge going to Trisul I. Thereafter the team was to move roped up in two sub teams, to make a direct attempt on Trisul II, from the northwest ridge of the mountain. Owing to unstable snow conditions, the traverse from Trisul II to III was ruled out.

In the late hours of 31 May and the early hours of 01 June, Nk Bishnu Bahadur, a strong member of the team started to show major symptoms of altitude sickness, which was later diagnosed as 'High Fever with Altered Sensorium', a very dangerous condition. He was delirious and fast slipping into a state of semi coma. As he became critical, the leader decided to abort the attempt on Trisul II and to evacuate Bishnu to Camp 2, at first light on 01 June. It was a crisis situation!

As Camp 2 was located on a sharp ridgeline, evacuation by air was not an option. Evacuation commenced at 07.00 a.m. on 1 June and was carried out with precision, over the exposed west ridge of Trisul I. The critically ill member was brought down over 700 m on a steep ridge, including pitches of blue ice of 60° slope by a team comprising of Subedar Topgay Bhutia, Assistant Leader Kunchok Tempa and Assistant Leader Tenzin Rigden supported by Brigadier Ashok Abbey and Rifleman Birendra Singh. The patient was successfully brought down to Camp 2 by 12.50 p.m. Subsequently, Captain Basant Pathak, the young expedition doctor, moved up to Camp 1 and evacuated the patient to base camp and below. The outstanding team work at Camp 3 ensured that the critically ill Bishnu Bahadur survived. This was in the finest traditions of mountaineering and the Indian Army.

Looking east from the summit ridge, L to R - Nanda Devi, Nanda Devi East, Devisthan I and II, Devi Mukut and Devtoli. In the foreground (right) Mrigthuni. (Brigadier Ashok Abbey)

Looking west - Nanda Ghunti in the foreground, with Kamet towering above. Hathi Parvat is prominently visible to the right. (Brigadier Ashok Abbey)

Despite the setback of a critical evacuation, but with the weather holding out, the leader decided on another summit bid to Trisul I. Accordingly, Team C comprising of Subedar Kunwar Singh, Capt Brijesh Kumar, Rifleman Pratap Singh and Rifleman Sunil Rai occupied Camp 3 on 2 June, for a second attempt on Trisul I. They were supported by Hav Matwar, Rifleman Yashpal Singh and Rifleman Min Bahadur Tamang at Camp 2.

The team left Camp 3 at 03.30 a.m. on 3 June 2013. Due to high winds, the team could not start earlier. Being extremely well acclimatised and hyper fit, the members moved swiftly. They reached the roping up point by 08.10 a.m. By 09.45 a.m. they reached the high point of the first summit party, but exercising caution they too did not tread on the heavily corniced summit top. The summit team under Subedar Kunwar Singh descended to Camp 2 at 02.00 p.m. on 3 June 2013, after winding up Camp 3. In a remarkable display of fitness and grit, the team descended, directly reaching the base camp by 07.10 p.m. the same day. The team had been on their feet for nearly fourteen and a half hours. Meanwhile, the support team at Camp 2 wound up the camp.

On 4 June, Camp 1 was also wound and the entire team finally concentrated at base camp the same day. They descended just in time, as there was heavy snowfall in the upper reaches from 5 June onwards. The expedition team finally reached Sitel on 10 June.

The team ensured that no non biodegradable waste generated by the expedition was left on the mountain. All returning members from Camp 1 and above and porters returning from base camp and subsequently ponies from Bhujani, were used for bringing down the non bio degradable waste generated from high camps to the base camp and then on to the road head. All members used a 'Kiwi Bag', which ensured that the mountain was not unduly littered.

In order to assist the civil population living in this remote area, two voluntary medical camps were held by the expedition team on their return, in the upper Nandakini valley. Approximately, 150 families of this far flung valley were given medical aid. The locals were extremely grateful to mountaineers and the Indian Army for this gesture.

The expedition team which reached Joshimath on 13 June, was coincidently present and thus naturally poised, when an unprecedented 'Cloud Burst' hit Uttarakhand on 16 June 2013. As most of the team members were former instructors and students of the High Altitude Warfare School, they immediately joined the rescue operations of the Indian Army. Thus the expedition team members not only spear-headed some these rescue operations, but rendered life saving aid, in which hundreds of stranded civilians were rescued.

The expedition to the Trisul massif in 2013 was a challenging and an eventful experience. While Trisul I had been climbed, the Trisul traverse could not be accomplished by the team as planned. However, the mountain environment had been left clean and pristine; medical aid was provided to inhabitants living in the remote, upper Nandakini valley; a fellow climber's life was saved by critical evacuation from summit camp on the west ridge. Most importantly, the expedition was on the forefront of rescue operations launched by the Indian Army, in arguably one of the worst natural disasters to hit the country.

> Somewhere, perhaps Trisul, Lord Shiva's omnipresent trident,
> had once again been the saviour!

Nomenclature in the Nandakini Valley

Trisul (Trishul)	-	Trident. It also reflects three humps on the summit
Bhujani	-	The place of *Bhojpatra* or the Himalyan birch
Bank	-	Glacier or body of ice
Dang	-	Big stone
Dudhang	-	Two big stones
Hom Kund	-	The place where 'havan' is performed in the memory of forefathers
Kund	-	Pond
Lata Kopri	-	Dead body
Nandakini	-	Water from the feet of Goddess Nanda
Nanda Ghunti	-	Face of Nanda Devi/ Face in a veil
Nanda	-	Golden Bhagwati
Kini	-	Water
Sili Samudra	-	Sea of stones
Tatra	-	Village of those who speak Tatri language

Summary

An attempt to traverse three summits of Trisul massif by an Indian Army team in May-June 2013. Trisul I was climbed on 30 May and 3 June, via the west ridge, by two teams.

5

First Crossing of Chaukhamba Col

A Historical Pilgrimage[1]

Debabrata Mukherjee

101 years ago, in July 1912, the legendary climber C. F. Meade reached the col between Chaukhamba I and Januhut, hiking from a high camp on Bhagirath Kharak glacier. He did not cross the col. He stood there for a while, got a glimpse of the Gangotri glacier on the other side, and then returned the way he came. In May 2013, 101 years later, Debabrata Mukherjee and his team climbed up the Bhagirath Kharak glacier to the Chaukhamba col and descended to the Gangotri glacier – a journey that began in Badrinath and culminated in Gangotri.

The first known crossing from Badrinath to Kedarnath was attempted by the legendary duo of Shipton and Tilman in 1934. Starting from Badrinath they followed the Satopanth Bamak to climb up the icefall of Chaukhamba I to reach the Satopanth col. Thereafter they were lost in the forest of Gondharpongi for ten days and couldn't find the way to Kedarnath.

Sixty-four years later, in 1998, Martin Moran completed the epic crossing[2] following Shipton's dream route, and reached Kedarnath following the upper reaches of the mountain.

In 2009 our team made the first opposite crossing - from Kedarnath to Badrinath - following the Mahapanth col, Vishali col and Yeonbuk col and completing the first encircling of Nilkantha and Parvati. However, it left me with a desire to complete the circle, crossing Chaukhamba col and Gangotri col, to reach Kedarnath.

It was towards the end of 2012 that I began my research on Chaukhamba col in earnest, which revealed many interesting facts. The first mention of the col is found in the well-known book *Kamet Conquered* by Frank

1. This expedition was awarded the 'Jagadish Nanavati Award for Excellence in Mountaineering' for the year 2013. This award is instituted by the Himalayan Club in association with the Nanavati family to encourage quality climbing, exploration and reporting by Indian mountaineers. - ed.

2. 'Shipton's Lost Valley' by Martin Moran, *HJ* Vol. 55, p. 63.

Chaukhamba col from Badrinath side. (Debabrata Mukherjee)

Smythe, in which he writes about C.F. Meade and his 1912 exploration of Bhagat (*Bhagirath*) Kharak gap at the head of the glacier with the same name. The exclusive pass is also written about in the book *Approach to the Hills* by C. F. Meade in 1932. Eric Shipton, in his book *Nanda Devi* (1936), wrote about his exploration of these watersheds before climbing Nanda Devi, and alludes to the pass.

It was surprising that despite its mention in three epic books, nobody had tried to climb or cross the Chaukhamba col during the 60 years following Shipton's failure on that route. The Kolkata section of the Himalayan Club reconnoitred the lower icefall below Chaukhamba I during their traverse of the Gangotri glacier in 1984. Later, a British team, under the leadership of Simon Yearsley reached the Chaukhamba col from the Gangotri side and camped there on 3 October, 1995 on their way to attempt Chaukhamba I, but found the next part of the route too dangerous to continue. The only attempt made to reach the col from the Badrinath side was by Harish Kapadia's team in June 1997[3], when two members and three Sherpas failed to reach the main glacier below the north face of Chaukhamba I.

3. 'Lost in the Legends' by Harish Kapadia, *HJ* Vol. 54, p. 73.

The significance of the Chaukhamba massif in the geography of India is unparalleled. It is the main donor to the Gangotri glacial complex which supports the water supply to the three main resources of the Ganga viz. Bhagirathi, Alakananda and Mandakini.

After some effort I got a determined team of four mountaineers - Biman Biswas (29), a combat commando of ITBP; Partha Sarathi Moulik (40), Biplob Baidya (42), and Ritabrata Saha (33). My trusted supporters Kamal Singh (40) and Devram (22) brought along two fellow villagers Kirtan (22) and Raghubeer (25) from Agoda (the second village on the trekking route to Dodital from Uttarkashi).

After several months of researching, we left Kolkata on 12 May, 2013 for Joshimath, where we spent two days getting our permits and shopping for provisions, and then reached Mana - the roadhead - on 16 May noon, bypassing Badrinath which had just opened its doors that day to pilgrims.

Camp I - Bagua Nala

We hit the trail on 18 May. The initial five km stretch, which leads to Vasudhara falls, is well-marked. After crossing the Bhagnyu nala (a stream coming down from the north), we reached our campsite (3528 m) in three and half hours.

Camp II - Chaukhamba I Base Camp

We had two options for our trek the next day: 1) a lower route - to follow the river bed initially over the dead moraine and then on the terminal moraine of the Bhagirath Kharak or 2) an upper route - go up to our right and follow the broken ridge created by the true left lateral moraine. We took the upper route. Crossing the boulder zone and once atop the lateral moraine, we found a pleasant trail between the moraine ridge and the northern wall. It took us three and half hours to reach the next campsite (Chaukhamba I base camp) at 3898 m. A small improvised temple indicated that we were indeed at a place used by the expeditions to Chaukhamba I.

Glacier Camp 1

On 20 May we started down a boulder gully, which was more difficult than we thought. The loose boulders on sand and scree were perched so precariously that each of us created a landslide towards the Bhagirath

L to R: Chaukhamba III and IV from Gangotri glacier. (Ritabrata Saha)

Kharak glacier. We followed an imaginary line to cross the glacier to its true right and soon found ourselves in a sea of boulders. However we managed to keep our sense of direction and reached the snowfield below the north face of Balakun in an hour or so. After a six hour trek we found a relatively flat place and lowered our rucksacks to set up our glacier Camp 1 at 4302 m.

Glacier Camp 2

The following day the gradient was constant at 35^0-40^0. With the rising sun we started sinking in the melting snow. As we crossed a turning on the glacier, the view of Chaukhamba I opened before us. The avalanche-prone north face of Chaukhamba was awe-inspiring, but the icefall below seemed a nightmare. Our goal - Chaukhamba col - was somewhere up and above that icefall. After a continuous trek of six hours on the crevasse ridden snowfield with knee-deep snow, we reached below the hidden gully which leads to the side glacier used by Meade's team in 1912 but was not found by Shipton in 1934. The low altitude supporters were sent back from here together with the stove, big utensils and other extra loads.

After setting up the camp we walked up to the base of the rock wall to assess the route ahead. Our first obstacle would be to climb the first 80 m of that 80^0-90^0 wall to reach the base of the snow couloirs. Snow avalanches were coming down every five minutes. We hoped there would be fewer avalanches in the early morning. Biman started fixing rope on the wall with pitons. A huge boulder on the top of the rock wall provided a safe anchor.

Phase II

The morning of 22 May ushered the second phase of our expedition - with three main variations from the earlier phase. Firstly, we would now be climbing, not walking. Secondly there were no longer any LAPs accompanying us, so we would have to carry and ferry the loads. And lastly, we would now be on a diet of soup and noodles!

Biman was the first to move out of the camp towards the base of the rock-wall. The subsequent critical part was to cross a snow-bridge over the funnel. The thin layer of snow over the waterfall was also the runnel for all the loose snow coming down every five minutes from the upper valley as mini avalanches. We couldn't fix any rope here. Instead, we belayed for a distance of 30 m. Biman fixed rope on the following 200 m - a 70^0 mixed patch of rock and snow. It was about 01.00 p.m., and the weather was at its best. So I decided to continue. Though we had run out of rope, I had a six mm 30 m line which we doubled and tied to our harness. After about two hours we reached the brim of the funnel and found a place which seemed safe from avalanches. Meanwhile, the other three members climbed up the fixed rope to the point where Kamal and others had left the first set of load. They were waiting for the HAS to get there with more ropes etc. It was already 03.00 p.m. and the snow was fast turning soft under the scorching sun. Tired and without any climbing rope, it could have been dangerous to continue on that slippery slope. I asked them over the walkie-talkie to stop and set up their tent. I knew it wouldn't be easy to place a tent on that 55^0 slope. Still it was a better option than continuing for another two hours in those conditions. They somehow dug out a spot for the tent and bivouacked for the night. Kamal and his team arrived there with the ropes after one and half hours but did not want to stay the night in the half-hanging position, so continued upwards leaving a climbing rope for the team members. Biman and I slowly unpacked, put up our small tent, brewed some soup and watched a fantastic sunset. We were soon joined by our HAS. The GPS reading showed the height as 5131 m.

Early next morning, while three HAS - Devram, Kirtan and Raghuveer - went down to bring the loads from the last bivouac, Biman and I, together with Kamal, moved forward. We crossed the avalanche debris and continued traversing upwards on the north face of the intermediate ridge towards the col some two km ahead and some 370 m higher. At about 11.30 a.m. we reached the intermediate col at 5490 m. Around that time the team members and supporters from the hanging bivouac arrived at our last night's campsite. While they stopped to repack and rest, we got our first view of our goal – Chaukhamba col. Facing south, from left to right we could see some of the stunning peaks - Mukut Parvat, Kamet, Mana Parvat, Nar Parvat, Narayan Parvat, Balakun and then Chaukhamba I (7138 m) with its crevasses and hanging glaciers and the continuous avalanches all along its north face. The summit looked so close yet so far with its two rock-slabs known as *charan-paduka* (slippers of Lord Vishnu). The northwest ridge continued from the summit to our right and suddenly dropped at the heavily corniced Chaukhamba col at 6053 m before rising again to meet the summit ridge of Januhut. Behind us were the Arwa group of peaks and many other known and unknown giants of Garhwal. The mere sight of the col sent a shiver down my spine. The glistening, curved icewall with the cornice on the top looked unachievable. Biman and I exchanged a glance and smiled. We went down about 60 m – or rather rolled over a crevasse to negotiate it and reach a flat spot. The walkie buzzed - only to deliver an undesirable message from the team below. Partha and other team members, sunk to their waist in the melting snow, were unable to move up with their already tired legs from the morning's climb plus lack of sleep , and were therefore returning to our last night's bivouac. So for the second consecutive night the team was in separate camp sites. We sent the four HAS to the lower camp. Biman and I sat at our camp site enjoying the view of Chaukhamba col and planning the route. The GPS registered this point as 5420 m.

We had a lazy morning at the upper camp on 24 May. When the rest of the team finally joined us at about 09.30 a.m. we had already anchored the only 220 m polypropylene rope we had, above a gap to the right of our campsite. We had to go down a couloir of 70^0 to reach the glacier under the north face of Chaukhamba I. By taking this route we could avoid the main icefall and the most avalanche-prone area but we would have to cross another two km of crevasse-filled glacier and risk avalanches from both north face of Chaukhamba I and south face of Januhut. Moreover we would have to cross the bergschrund of the Januhut-side several times - and reach a safe site before dark. Where, when and how - we still had to figure out.

The 250 m descent was uneventful except for Partha slipping and somersaulting. Once on the glacier, we roped up in three groups. We tried to stay away from the already melted snow, keeping close to the south face of Januhut - a perilous task with hidden crevasses every five steps, and avalanches of different dimensions every five minutes. It was an unbroken upward movement for the next four hours, without any substantial rest. At about 03.00 p.m, I reached the last section of the glacier, 500 m before the final wall of Chaukhamba col. Devoid of sunlight, it suddenly felt very cold. We called it a day and started pitching our tents. We slept to the music of avalanches at a height of 5623 m.

The next day we started towards the col keeping close to the northern wall. There were two lines of bergschrund to cross in the beginning of the climb. We climbed the first one over a thin snow-bridge and avoided the second by climbing up a small rock projection. Continuing to climb an even steeper slope after those crevasses, we stopped near the next rock island. We had already climbed for two and half hours from camp. Dumping our equipment in a snow-hole, we returned.

26 May – our D-Day! So far we had enjoyed unbelievably good weather. But on the morning of 26 May, the sun rose behind a curtain of heavy mist. We left our tents at 06.30 a.m., laden with heavy loads, to make the historic climb to the col. Charles Francis Meade climbed it with his Swiss guides, Pierre Blanc and Justin Blanc (two brothers) and three Bhotia porters, on 10 July, 1912. They had perhaps less snow cover over the rocks and more snow on the glacier and ice wall – it was a less warmed up planet. A few lines from his book must be quoted here to understand the condition of the wall 101 years ago.

> Above this bergschrund the angle steepened so that in some seasons the snow might be avalanchy and dangerous, perhaps forcing a climbing-party to take to some rock-cliffs on the right. As usual in the Himalaya, the angle turned out to be much steeper than a distant view of it had led us to suppose, but with increasing eagerness to see the unknown country ahead, we raced panting up the final slope. Suddenly we found ourselves on the top of the pass, and a thrilling spectacle was revealed.

There was no mention of the cornice on the top. While before us was an overhanging 50 m icewall with a three m hanging cornice! And it was very avalanchy in May. We began climbing the rock-cliff, but the thin snow cover over this 80^0 rock wall made it difficult to keep our feet stable. We were moving in our usual groups, and reached the spot where we had dumped our load the previous day. After picking up the load we started to crampon up the steep slope. Every step was a series of actions; first to determine the safety, then stability of the snow / ice, then balance of the body, then decide the direction of the next step and finally raise one foot to place it a little higher up and start panting for some oxygen. We forgot how long we moved in this manner. Suddenly I found that I could see the unnamed peak attached to Chaukhamba IV. The cornice of Chaukhamba col was about 70 m below. We crossed the ridge and entered the saddle connecting Chaukhamba I and Januhut. We were at Chaukhamba col!!! Biman and I quickly congratulated the supporters, before trying to shelter ourselves, behind some rocks, from the high winds from the Gangotri glacier. Before we could unbuckle our rucksacks, Biplab's SOS came through the radio. Partha had slipped and was somehow arrested by Ritabrata and Biplab from going down 300 m into the gaping bergschrund. Still shaken from the fall, he couldn't stand up and place his steps properly. They needed help with Partha's load. We rushed towards the edge of the col to find, about 100 m below us, Ritabrata and Biplab waiting, anchored with their ice-axes while Partha was hanging between them. Our three HAS rappelled down to help Partha. Kamal took Partha's rucksack and tried to clip his jumar to the fixed rope. Suddenly Kamal started to skid - first slipping down, then rolling and soon bouncing. We shouted out to him, 'Throw the rucksack away!' After 150 m or so, the rucksack released, and Kamal managed to dig his crampon and hands into the snow and stopped his fall. Partha's rucksack continued bouncing for another 150 m and luckily stopped just before an open crevasse. Kirtan and Raghuveer went down all the way to the crevasse with the short climbing rope, recovered the rucksack and came up after about two hours.

Finally when they came up we took photographs and had a small ritual. I tried to locate the cairn placed by Meade but couldn't find anything after 101 years. At about 02.30 p.m. we started to move down the other side. The weather felt strange. A dark cloud over Chaukhamba massif was trying to spread over Gangotri glacier but the strong wind (over 140 km/hr) was pushing it back, keeping the sky clear along the glacier. It was an amazing view. Once we got down to the saddle, we had to make a human chain, because of the strong wind, to cross a 40 m open passage leading to the other side of the gap. After traversing the saddle towards southwest we began our descent, relying on the contour map and Google Satellite images. We subsequently found that GE images were not correct for this part.

After some 100 m downward walk, we found that the snow slope had suddenly changed to rock cliffs on all sides. We couldn't see anything below. After a discussion, we anchored two snow-stakes and Biman started to descend through a snow-rock couloir. Devram, Kamal, and Ritabrata went down to help him fix the two climbing ropes. The rope already had two knots, so it was difficult to change the descender twice in the middle. Partha, still very much shaken from the fall, took about an hour to go down that first stretch.

With sunset approaching, the wind started to blow powdery snow. It became increasingly difficult to stand with the heavy rucksack over the anchor and wait for the last man to reach the next anchor. Once everybody was down, it was my turn to descend. I had supposed that all the knots were on the same side of the doubled rope and therefore I wouldn't have any problem to pull it down from the other side. But to my bad luck I found they were on either side of the strands. So I had no other option than to untie the knot and climb down unprotected on a steep ice slope. I was just about 30 m from the next anchor when I slipped. However, I kicked with all my strength while skidding and somehow my crampons got stuck.

Darkness fell over Gangotri glacier and soon over Chaukhamba massif. We continued our downward journey in the glow of light from our headlamps. At about 08.30 p.m., Biman reached a flat spot and started pitching tents. I could see the distant light of the tents when I heard Partha moaning in pain, fighting to come out of the last bergschrund. He had torn his ligament when he fell in the first bergschrund and twisted his legs. However, he managed to come out and started towards the campsite. When I reached the campsite it was 11.00 p.m. Seventeen hours of continuous climbing on an unknown route got us mentally and physically tired. It was impossible to sleep that night with five of us crammed up in one tent. However, we had crossed the Chaukhamba col and were now at 5563 m.

The Descent

We began our descent on the morning of 27 May, with Partha in pain. He couldn't stand properly. Not long after we were on the top of the Chaukhamba I icefall of Gangotri glacier. To our right was a rock wall dropping straight into the glacier and to our left were the menacing seracs of the icefall. We had no option other than taking the chute to go down. We went down some 50 m on easier gradient and placed the anchor for the last possible abseil to reach the Gangotri glacier. Meanwhile the weather packed up and it started to snow amidst high winds. Biman continued to rappel down and found there was no straight route downwards, so he zigzagged and after two intermediate anchors hit the glacier floor after more than one hour. By this time the blizzard had gained strength. The superfine powdery snow was lashing at great speed through the one metre narrow chute and entering our nose, eyes,

Route of descent through the ice-rock gulley. (Debabrata Mukherjee)

and ears. The numbing cold wind was freezing our joints and limbs. One by one the members started to go down. After about six hours it was my turn to descend, followed by Devram. When I finally stood on Gangotri glacier, golden sunrays were just touching the tip of Chaukhamba I. I took some hurried photos of Devram coming down the last few metres before setting up my tent (5245 m) with an uneasy feeling of uncomfortable closeness to the hanging seracs of Chaukhamba I icefall and the continuous small avalanches from Chaukhamba IV. We made some tea and had barely had a sip when an unusually big boom of an avalanche sounded and Devram shouted, 'It's a big one coming down - it's coming to our tent.' We ran out and saw a greyish white 100 m high wall of clouds thundering towards us. Biman shouted, 'Get into the tent and put your back to the tent wall towards the avalanche.' We followed his command like puppets. For thirty long seconds there was violent pushing and shaking and then suddenly everything quietened. We came out of our tent and saw how lucky we were - the snow and ice debris had stopped about some 100 m before the camp site - what we felt was the wind-thrust and snow-blast.

Once again a bright morning welcomed us. I wanted to leave as soon as possible from the dangerous seracs around us. But Partha reported that he couldn't raise his leg more than six inches; Biplab informed us that the butane gas we were carrying would not last longer than two days.

Chaukhamba col from Gangotri side. (Debabrata Mukherjee)

View towards Gaumukh. Mandani Parvat and Yeonbuk in the background. (Debabrata Mukherjee)

Our planned route would take us at least five days to reach Kedarnath via Gangotri col on the western boundary of Gangotri glacier. We, therefore, decided not to try the Gangotri col. Instead, we started walking on Gangotri glacier towards Gangotri. After walking for five hours we stood facing the Kharcha Kund peak and stopped for the day at 4691 m. Partha arrived, escorted by Raghuveer, two hours later.

The next day we got down to 4412 m at the junction of Kirti and Gangotri glaciers near the turning for Kedardome base camp. It took us four hours to get there.

We left the following day for Tapovan. With all the known peaks around us - Kedardome, Kedarnath, Shivling and Bhagirathi - it felt like a familiar neighbourhood.

On 31 May, we completed our pilgrimage at Gangotri temple.

Summary

First crossing of the Chaukhamba col (6053 m) linking Bhagirath Kharak and Gangotri glaciers in May 2013.

6

Exploring the Valley of the *Jadhs*

The Jadung Exploration, June 2013

Ashutosh Mishra

'They must have cooked the name up' I thought when one of the officers of the *Jadung* post said... 'Sure, ten km further up there is a beautiful lake – Janak tal.'

After names such as *Jangla*, *Kopang, Karcha, Nelong, Sumla, Dumku, Pulamsumda*, names that had such apparent Tibetan influence, the starkly Hindu name was a surprise; *Janak* - the *Sage-King* and famed father of *Seeta* in the epic *Ramayana*, one who finds several mentions in ancient Indian *Vedic* texts such as the *Upanishads*! We had just reached the Jadung post in the central Nelang watershed, after a 50 km drive from Bhaironghati, along a road off the national highway to Gangotri.

A month later, scanning again through the research material on the expedition we had just finished, I re-read a long transcript of an

Janak tal. (Ravindra Bhatt)

interview[1]. The interview was with an old Jadh lady - Shanti Devi who seemed to be happily recalling the times when Nelang and Jadung had seen better days. I remembered this document because this was the only document we had found before the expedition that had any description of the terrain ahead of the Jadung village. After reading the interview, I realised, there indeed was a local legend about King Janak. The Jadh villagers of Jadung perhaps used the name Janak tal for many centuries. The ITBP had evidently not cooked up the name of that pretty jewel of a lake.

The Jadhs of Jadhung (Jadh-Dung)

The name Jadung seems to have come from Jadh (name of the tribe that inhabited the valley) and Dung (mountain). The British mapmakers have mentioned this place as Jadhang, but all local residents of Uttarkashi use the name Jadung (perhaps Jadh-Dung) for this village and valley. The pronunciation agrees well with the nomenclature. Agriculture, livestock and commerce drove economy of their villages of Jadh-Dung and Nelang[2]. The Jadh population[3] is estimated at around 2500 today of which more than 60% are residents of the Kinnaur valley in Himachal and the rest in the villages of Harsil and Dunda in Uttarakhand. Although of predominantly Bhotia origins, the Jadh people have a strong Garhwali influence. Even though they have converted to Buddhism and changed their rituals during the time of Tibetan domination; the legend of King Janak and festivals like *Pandav Lila*[4] continues to live on in the Jadh consciousness.

Although the only source of documented history of the Jadh people and region is the archives of the states of Garhwal and Bushahr as well as writings of European explorers, none of the documents can be traced back earlier than 1700s. Europeans started taking interest area from 1807 onwards.

Pre-British Era (*c.* 16th to 18th Century AD)

The entire area at the source of the Bhagirathi including the Nelang watershed remained mysterious and obscure to the native Indian as well as the British for a long time because of its inaccessibility. The tall walls of the Nelang gorge at Bhaironghati barred the way of the most intrepid traveller.

The easiest access into the Nelang watershed was from Tibet over the Tsangchok la or Jelu Khaga. This may account for the influence of the Buddhist religion and customs that may have proliferated from Tibet[5] in the late 1600s and early 1700s during the time of the 6th Dalai Lama.

It seems, the Jadhs used to migrate downstream of Bhagirathi and Baspa for trading in winters, thus inviting the right of revenue taxation by the states of Garhwal and Bushahr respectively. Similarly in summers, they had easy access to Chaprung of Tibet where they used to trade in Indian goods, thus inviting taxation from the Dzongpen of Chaprung! There is apparently a copper inscription of 1667 AD that states clearly that the Kingdom of Bushahr has gifted the Nelang tract to the Tehri state[6].

As late as 1800 AD, armed cavalry from Bushahr state have been said to cross into the Nelang area over a pass called Chunsa-Khaga[7] to collect taxes and dues[8]. State archives of the 1700s indicate rift between the princely states of Bushahr and Garhwal regarding matters relating to collection of taxes and tributes[9] from the Nelang tract. It is apparent from many other accounts that, in the earlier part of the 1800s, the villages of the Jahnavi watershed had a dual identity by the name of 'Chounsah'[10] as called by the Tibetans and 'Nelang' and 'Jadung' as called by the people of Bushahr and Garhwal. As early as 1804, the Nepalese attacked and destroyed the Nelang village.

The British Era

Just after the Anglo-Nepalese war was over in 1815 the political boundaries in Garhwal was redefined and Garhwal was divided into two parts. The British Garhwal (Pauri and Alaknanda valley) was established and the rest of Garhwal was consolidated under the Tehri-Garhwal state. King Sudarshan Shah, the heir of the earlier King Pradhuman Shah, was reinstated as the ruler of the new state. With the support of the British, the King ruled for 44 long years[11] ushering in stable governance for the state.

J B Fraser visited the Upper Taknore valley around 1815 in his search for the source of the Ganga and documented the existence of Chaunsah or Nelang and wrote about the Jadh Bhotias. He recorded the winter migration of the Jadhs into the Harsil area for the first time. Captain John Hodgson and Lt James D Herbert of the Survey of India carried on from where Captains Webb and Raper had left in 1808[12]. Hodgson became the first European ever to visit Gaumukh on 31 May 1817.

Lt Herbert returned to the Upper Taknore valley in September 1819[13]. He became the first European to traverse up the Jahnavi river and reached Nelang on 13 September 1819. He mentions that the Jadh people of Nelang and Jadung were apprehensive of European incursion and discouraged him from exploring till the pass to Tibet. The British suspected that the pass of Tsangchok la (Jelu Khaga/ Jeela Kanta)[14] in Nelang valley could be an easy gateway for attack by the Russians. They needed someone nearby to be their eyes and the ears[15] in this region.

During the 1840-83, the Harsil valley saw the emergence of Fredrick Wilson – a British army officer who deserted the army and chose a life of enterprise. The British found a convenient point man in him for gathering crucial intelligence about the happenings at the Nelang border. In his early days at Harsil, Wilson, in 1842, followed two Russian army officers through the Chor gad valley of Nelang, over the Gundar pass[16] (Perhaps a Garhwali name for the pass that Gerard mentions as Chunsa-Khaga) and reached the Bushahri village of Chhitkul. Similarly, in July of 1845, Wilson played a pivotal role in preventing the party of Prince Waldamer of Prussia to enter the Nelang valley[17].

Wilson was later appointed as the agent of the King of Tehri state and was also given the responsibility of rehabilitating the Upper Taknore and Nelang area in 1849[18]. He got few Bhotias from the Baspa valley to settle at Nelang[19]. The Tehri state played active role in rebuilding the village of and for establishing the hamlet called Jadung, deep inside the Jahnavi watershed[20]. Wilson was in touch[21] with the Jadh Bhotias on a regular basis[22].

C L Griesbach from the Geological Survey of India visited the Nelang valley in two successive years 1882 and 1883[23] and published his exploits in 1891. He made a detailed study of the geology of the region and documented the Jadh people and their valley[24]. Marco Pallis in his famous exploration[25] up the Gangotri valley in 1933, mentions his intent of hiring Jadh Bhotias as porters during his brief halt at Harsil. In the 1936-40 period the entire Nelang watershed was mapped with modern methods with people such as Lt J F S Ottley[26] and J B Auden[27] playing crucial roles.

Current Affairs

In April 1956, an armed detachment of the Chinese army came up to a mile east of Nelang village[28]. The presence of the Indian armed forces in the village increased steadily after the Chinese aggression of 1962 till the villagers were completely evacuated in 1965. Those Jadh villagers were subsequently settled at two places - Bagori near Harsil (summer) and Dunda near Uttarkashi (winter).

Romesh Bhattacharjee and Harish Kapadia visited the area in 1990 and documented the Tirpani gad and Mana gad valleys respectively. Today, nestled in that pretty valley of Jahnavi, the quaint little village of Jadung stands neglected and ruined. However, during all these developments, geographical details of the Jadung area were never recorded. In fact most of the mapping and documentation focus of the British explorers were on the passes going out to Tibet or Badrinath - (Tsangchok la/ Jelu Khaga, Thaga la and Muling la to Tibet and Tara col / Arwa col to Badrinath). For example the British army map of Garhwal in 1936 does not mention Janak tal ahead of Jadung.

Expedition Team

We were almost the same team that had explored Auden's Trail the previous year[29] in the same watershed. Sanjit Bal and his wife – Atamjot (a doctor) were new members. Apart from being regular 'crossfitters', they were recommended by Anant Khirbat (a key member of the team). Finally by early April 2013 we were a team of nine - Anant, Arun, Atamjot, Bharat, Kalyani, Kuntal, Ravin, Sanjit and I.

Topographical Backdrop and Route Plan

The Nelang watershed drains water from four major valleys. Three of these vales have a generally north-south orientation and only one, the Mana gad, has an east-west orientation. The three valleys that have the International border to the north and the Jadh ganga to the south are: Chor gad valley in extreme west, Tirpani gad valley in the extreme east and the Jadung gad valley, in the middle. The Tirpani gad, Chor gad and Mana gad valley have seen trade traffic since centuries. The history of these vales is well documented by Atkinson, Heinrich Harrer, Ottley, Auden[30], Romesh Bhattacharjee[31], Harish Kapadia[32] and Tapan Pandit.[33]

Jadung valley bears little geographical description in historical records especially about details of what lies between Jadung and the bounding

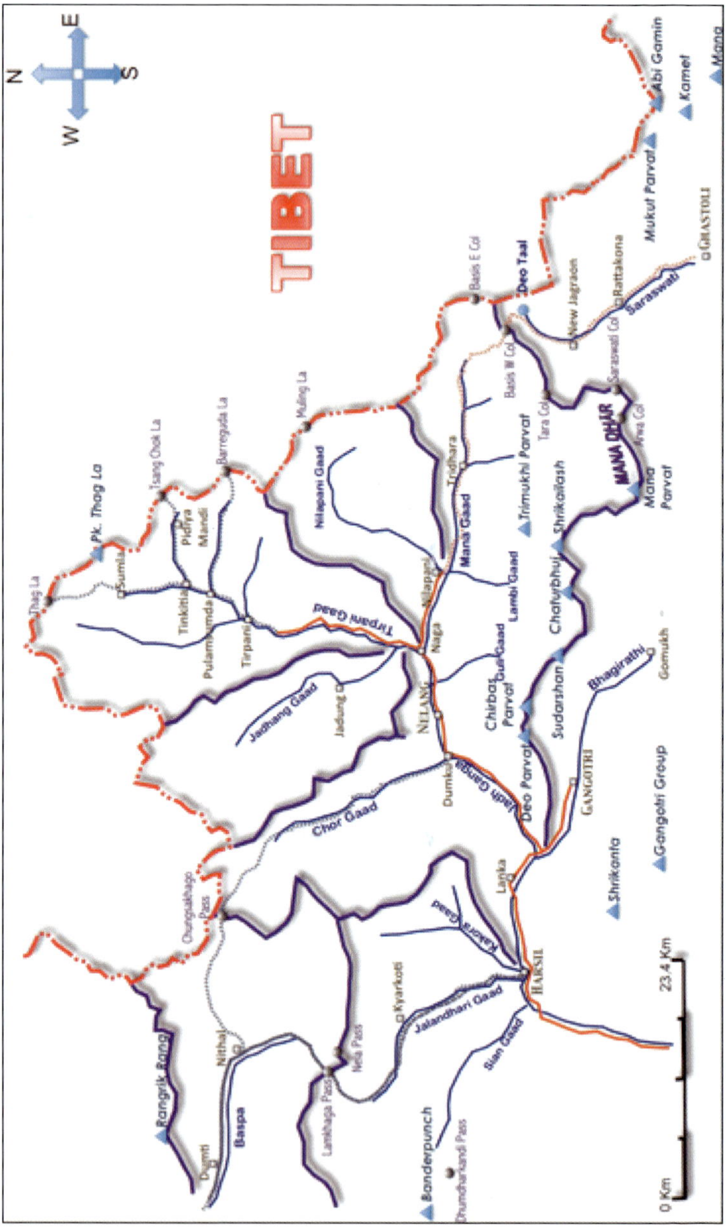

Jadung Area Map.

ridges of the water parting lines 20 km to its north. In the age of Google Earth and NASA Worldwind it is possible to have a look at the geographical features in a 3D model of the globe. We utilised this modern-day technical advantage to the hilt while planning the route.

Route Plan

Our initial research revealed a beautiful blue lake roughly 150 X 50 m in size, about 10 km ahead of Jadung village. In the satellite images, one could clearly see a well laid out track leading all the way to the lake and then on along the Jadung gad for a few kms. But it looked difficult to navigate northward along the main valley all the way to the headwaters. However, hope beaconed to the true right of the valley near the lake where a dead glacier appeared leading up to a high plateau ruled by a beautiful looking peak. The peak had shoulders on to its north and south both of which seemed to have doable cols that could lead one to the neighbouring valley to the west, the Chor gad valley.

We planned to explore this glacier, crossover to Chor gad using one of the cols and then proceed north to the head of the Chor gad, crossover the pass Chunsa Khago to the Baspa valley and exit at Chhitkul in Himachal Pradesh. We aimed to execute the plan in the first half of June and exit to Chhitkul on 17 June before the weather really worsened.

Phase 01- Walking in the Jadung Valley

Uttarkashi (1100 m) -125 km/6Hrs - Jadung (3650 m) - 7.5 km/6Hrs - Bharal Kill Point (3950 m) - 2.5 Km/2 Hrs - Janak tal (4120 m)

With a mild drizzle providing wet cover, we started on 02 June at about 9 a.m. and after a pleasant drive we were at the gateway of the Nelang valley by 02.00 p.m. After some minor formalities with the post-commander at Nelang we proceeded forth, seeing Naga and Do Sindhu bridge en route. Finally when we reached Jadung around 03.00 p.m., it was time to negotiate another administrative hurdle.

Due to some communication issue, the post-commander at Jadung had not been informed about our arrival through official channels. By about 8 in the evening the official hassles were sorted out. The day ended with a nice campfire in the farmlands of Jadung. The little solar-powered lights of the ITBP post shone at a distance across a little stream. The silhouettes of the ruins of the Jadung village around created an ambience that was surreal and ghostly.

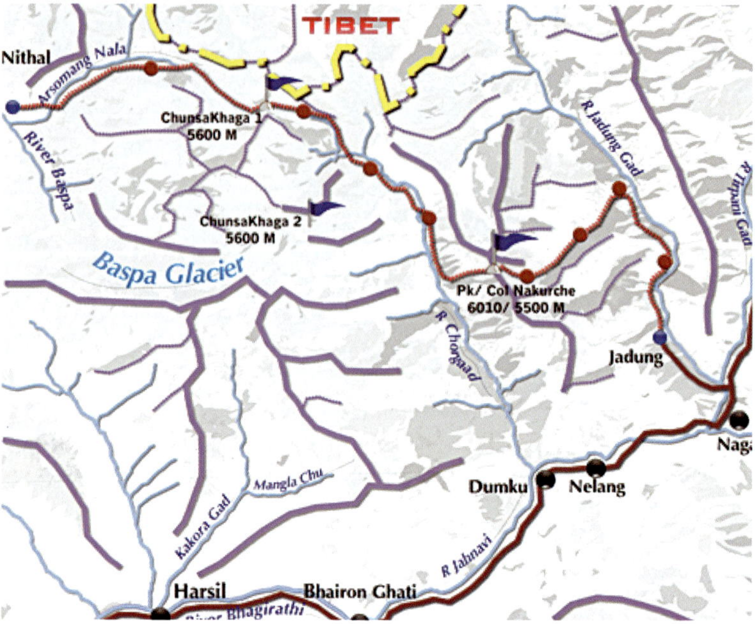

Trek route.

The next day, we arrived at the lovely camping grounds of Lal Devta just within an hour of starting off from Jadung. Here, the Jadung gad curves in from the west with a wide bed and a giant bend. The campsite has verdant terraced plains with little shrubs of yellow flowers. Almost in the middle of the ground there is an ancient place of worship and nearby are remnants of shepherd shelters.

The route rises steadily thereafter to a raised plateau, well crafted into terraced fields. Towards the east almost on the high right bank of Jadung gad is a small temple with a timeworn brass bell, dedicated to Lal Devta - a deity traditionally revered by the Jadhs. Ahead was a glen with a fast-flowing stream, beyond which the route rose sharply leading to another rocky shrine atop a spur with a red pennant atop fluttering capriciously. A small group of ITBP men were huddled around that shrine as we could see from a distance. They had gone to dump advance load for an exercise that was to take place in a few days in the area ahead of Janak tal.

The route ahead eased out. We were now on a valley terrace, which was a few hundred metres wide with a sharp rise to the ridges to our left and

A house in Jadung village. (Sketch by Kalyani Patil)

a deep dive into the riverbed flowing to our right. Kuntal and Anant were the advanced scouting party and they reported a possible campsite by the riverside. I estimated the lake to be about three km away. The campsite looked good for it allowed less distance to be covered the next morning.

As the team approached the campsite a ghastly scene awaited us. Right in the middle of the identified camping area, there was a carcass of a dead Bharal. One side of its face was eaten up and there were signs of scavenging on his stomach. It appeared as if the animal was killed by a leopard but had been eaten up partially either by a pack of dogs or by a bear. One thing was clear. More than one predator ruled the area. We took extra care that night to discourage local predators by lighting up small fires. Next day, after a sharp climb from the Bharal camp, the valley opened out and the route eased off. Within an hour we reached a vast camping ground. The GPS indicated that Janak tal was still half a kilometre away. We called the place 'Janak tal - Shepherd camp'.

At 11.00 a.m. we entered an amphitheatre with a spectacular setting. The Jadung gad tumbled down furiously from the north in easy steps from between narrow vertical walls of the gorge ahead. A clear blue

Looking back to Jadung from Lal Devta grounds. (Kuntal Joisher)

stream from the true right came in and met with the Jadung river in this open area. As we proceeded further we could see the origin of the clear-blue stream. There it lay, nestled cozily between two mountainsides, the aquamarine blob of nature's marvel - 'the Janak tal'. When we saw it for the first time, we were wonderstruck by the vibrant blue of the Janak tal, as if someone had accidently constructed an Olympic-sized swimming pool in the wrong place.

We noticed a pair of dark objects swimming in the middle of the lake. It was a pair of 'Northern Pintails'- a species of goose that's found over a large area in the northern hemisphere. The strange thing was, they were the only sign of life in that secluded piece of heaven. The lake Janak tal has an oblong layout and is set on a southwest-northeast orientation. The southwest tip of the lake is adorned with the terminal moraine of a dead glacier. This glacier, our object of exploration, was christened the 'Janak glacier'. From here onwards our path departed from the ITBP patrolling route, up through the rocky bed of this glacier.

A recce team left to explore the route ahead and identify a campsite for the next day. Near the lake we found an old little temple with figures of goddess Durga. Nearby was a pile of flat rocks upon which were

Janak tal surroundings. (Kuntal Joisher)

inscribed some prayers/writings in the Tibetan script. We identified some of the inscriptions saying *Om Manipadme ham Namah-* The famous Buddhist prayer. These were few mani stones that some of the British explorers had written about. The proximity of the two places of worship indicated religious influences that would have impacted the Jadh people in the times gone by. Since Janak tal was deeply revered by the Jadhs, it appeared the Buddhists from Tibet also wanted to leave a mark here.

The recce team came back at 03.00 p.m. Their report wasn't encouraging. They had climbed about five km till a rocky flat. But they had not found any water source. Apparently the entire glacier was a dry rocky desert. We hoped that there would be patches of snow, which we could possibly melt.

Phase 02- Exploring the Janak Glacier

Janak tal (4120 m) - 5 km/5 hrs - Tower Peak Camp (4600 m) - 3 km/ 3hrs - Jadung kund (5050 m) - 2 km/2hrs - ABC (5300 m) - South col (5500 m)/ Summit ridge (5850 m) - 10 km/5 hrs - Janak tal (4120 m)

The route to Janak glacier climbs steadily due southwest from the extreme tip of the Janak tal. Soon we were deep in the middle of brown rock country. Tall craggy ridges capped with melting snow flank the moraines of the Janak glacier. Soon we reached the site where loads had been dumped the previous afternoon. We still had not found water!

Fortunately, Kuntal and Anant struck water just about 200 m ahead on the true left of the medial moraine. We established our first rocky camp. We had made good progress for the day. The altimeter read 4600

Tower peak seen from Janak glacier. (Kuntal Joisher)

m. On to the true right of the glacier just opposite to our campsite rose a tall rocky tower capped with snow, looking like a smaller replica of Kailash.

Continuing on a similar terrain the next day we proceeded for two km to reach the confluence of the glaciers coming from north and south of the Nakurche peak. Our water woes were over for we noticed a multitude of glacial pools all along the way. The way ahead was now snow covered with a thin rivulet of snowmelt stringing through it. About a kilometre west of the junction we came across a large glacial pool, which we called as Jadung kund, located close again to the true left of the glacier. We had already crossed the 5000 m mark. It was time to set up our first snow camp!

Our earlier plan was to attempt a peak from this camp using the southeast ridge. But the route did not look very friendly from where we were camped. We decided to set up an advanced base camp at the cwm of the glacier from where we could carry out a simultaneous reconnaissance of both the Nakurche south col as well as the Nakurche peak. The gently rising snowfields led us to the southern head of the Janak glacier the next morning. After a friendly snow trudge of about two hours we reached a half frozen large glacial lake at 5300 m, just

Route ahead of Jadung kund camp to the head of Janak glacier. (Ashutosh Mishra)

100 m short of the bounding northern wall of the glacier on top of which sat the Nakurche peak. We called the lake Jadung tal and found a rocky patch on its high bank to set up camp.

Facing towards the head of the valley, the south col and the Nakurche peak were almost directly overhead to our left and right respectively. The south col and the route to its top looked dangerously steep. Towards our right, the route to the top of the summit ridge of Nakurche did not look easy either. It was already 7 June. Bad weather was closing in fast and we had to be off this high place within the next three days, either towards Chor gad, over the south col or back the way we came.

The quartet of Vinod, Janak, Lakhpa and Rajender left for a reconnaissance early next day, which opened sunny and bright; Vinod and Rajender went for the col and the other two for the summit ridge. After about an hour the party for the col led by Vinod radioed back news of their successful climb to the top of the col and reported the conditions for descent to Chor gad. According to them it

Nakurche peak from Janak glacier. (Ashutosh Mishra)

was dangerous to descend the almost vertical loose rocky slopes on the other side, particularly for the porters. There was no point risking an unsafe descent when the weather was closing in. I decided to attempt the peak and then retreat back to Jadung.

Janak and Lakhpa, the scouts for the summit ridge, returned three hours later just when the clouds were rolling in. They reported good news. Excitement was back in the camp. Everyone was soon busy rounding up personal equipment.

The Climb: ABC (5300 m): Rocky Pinnacle - Summit Ridge 5852 m, 8 hrs

We started at 04.00 a.m. The climb was hard and became risky when loose rocks started rolling down just when we were about 200 m short of the crest of the summit ridge. It took a steady climb of three hours to reach the ridge top to have a magnificent spectacle of the entire Garhwal Himalaya laid out before us. As we looked down along the ridge we could see the hump of 5750 m at one extreme - the point that's visible above the Jadung kund camp. The top of it looked like a helipad! Looking up we could see the gently rising but highly rocky ridge leading to the peak.

The walk on the summit ridge of Nakurche is a painful one. One has to continuously climb up and down the many small pinnacles that dot the summit ridge with a view of steep slopes leading down on both sides. As we trudged ahead the north col came into view to our right. This one looked perfectly do-able! I regretted having chosen the south col for our attempt.

After about an hour of arduous labour we reached a tall rocky pinnacle of about 30 m height. Only a traverse on the lower level looked feasible. It was already 08.30 a.m. and any such detour would add at least three to four hours, I calculated. The weather was predicted to be nasty after 12 noon. I had to take a call. 'We go down guys. Perhaps some other time!' I told the team. We were about 160 m below the summit and perhaps less than 300 m away. As one looked along the ridge ahead, the pass over northern Chor gad glacier to Baspa valley was visible. We had originally intended to crossover to Chhitkul had we been successful in crossing over the south col. All along the right was the boundary of the Nelang watershed that gave birth to the Jahnavi. At 2 o' clock direction one could see the headwaters of Jadung gad, a highly inaccessible and desolate place strewn with crisscrosses of myriad glaciers. Slightly to

View towards west from Nakurche South col. The mountain barrier defines the boundary of Chorgad valley. (Vinod Panwar)

its right were the extreme northeastern borders of Nelang- Thaga la and Tsangchok la were barely visible.

Straight-ahead, between Chor gad glacier and the Jadung headwaters the giants of Himachal - Leo Pargial and Reo Purgial were visible and so was Rangrik Rang. To the right and till the 4 o' clock direction our view was blocked by a massive rocky mountain adorned with many rocky spires. To all of our left the view was mesmerising. The whole set of ranges bounding the Upper Taknore Patti were visible- Shrikanta, Gangotri, Auden's Col, Jogin, Thalay Sagar, Bhrigupanth, Shivling, Meru, Satopanth, Chirbas, Deo Parvat, Sri Kailash and far in the distance the Zaskar giants Kamet and Mana.

The way down required a bit of rappelling and lowering down. All of us reached camp safely by 12 noon. By evening it was clear that the weather was indeed turning for the worse. On our way back to Janak tal, we observed a bright round halo of a rainbow around the sun. Only later we realised the significance of it when the freak cloudburst over Kedarnath on 15 June created an unprecedented and unfortunate disaster.

The Jadung gad had swelled up compared to what we had seen on our way up. There were more evidences of flooding on our way back to Jadung next day. The trail had gone underwater at several places. We spent some time at Lal Devta. It seems that it has always been an important deity worshipped by the Jadhs. Even near Bagori, where the Jadhs live today, they have created a shrine for Lal Devta, which one finds on the way up the Jalandhari gad valley. It seems thus that the Lal Devta shrine of Jadung was the original shrine of worship, which the Jadhs may have used for many centuries.

We reached Jadung Post in another couple of hours. Next morning we had a lovely walk roaming through the ruins of the abandoned village of Jadung. There are 30 odd houses in the village. The remains of the life of a people who used to live there could almost be smelt and touched. The pillars with fine woodwork, the cold stoves and smokeless chimneys and the empty granaries told a story of a time gone by. On the way we could see the Bhagirathi angry and swollen. May be it was a condition of extreme snowmelt and some heavy rains upstream. Luckily the vehicle arrived and nothing untoward happened on the way.

Two days later the news trickled in about the mammoth cloudburst over Kedarnath and the subsequent tales of devastation. One can only hope that life returns to normal as soon as possible and the economy of the mountain primes up again, very soon; but perhaps in a lot more considerate way for the Grand Old Man - the Himalaya.

Summary:

Exploration of Jadung valley, Janak tal and Janak glacier in the Jadh ganga valley of Garhwal Himalaya during June 2013.

---

1. 'Interview with Shanti Devi', *Mountain Voices – India- Garhwal* and *Kinnaur http://mountainvoices.org/Testimony.asp%3Fid=38.html*
2. P. 336, Vol III, Part I and P. 638-639, Vol III Part II, by E Atkinson, *Atkinson's Gazetteer-1882*
3. 'The Jads of India':http://www.joshuaproject.net/people-profile.php?peo3=16973&rog3=IN
4. 'The Descent of Pandavas-Rituals and Cosmology of the Jads of Garhwal' by Subhadra Mitra Channa, *European Bulletin of Himalayan Research 28: 67-87 (2005)*
5. Page 63 Para 1, *'Report of the Officials of Government of India and the People's Republic of China on the boundary Question- Part 3'.*
6. P. 49, *'Notes, memoranda and letters exchanged and agreements signed between the governments of India and China- 1954-59',* White Paper by Ministry of External Affairs, Government of India.

7. P. 150, Para 3, by Wilson 'Mountaineer', *A Summer Ramble in the Himalayas-1860.*
8. P. 49, Para 2, by Alexander Gerard, *An account of Koonawar- 1820.*
9. P. 54, Para 1, *Report of the Officials of Government of India and the People's Republic of China on the boundary Question- Part 3- 1960.*
10. P. 480, Para 3, by JB Fraser, *Journal of a tour through part of the snowy range of the Himala Mountains and to the source of the Rivers Jumna and Ganges- 1820.*
11. Tehri Garhwal- Brief History: *http://www.royalark.net/India/tehri.htm*
12. P. 33, Para 13-14, 'The Mountain Provinces- Garhwal and Sirmor' by Col R Phillimore, *Historical Records of the Survey of India,* Vol 3-1954.
13. P. 36, P.37, Para 1, 'The Mountain Provinces- Garhwal and Sirmor' by Col R Phillimore, *Historical Records of the Survey of India, Vol 3-1954.*
14. P. 148, Para 1, by Wilson 'Mountaineer, *A Summer Ramble in the Himalayas'-1860.*
15. P. 25-26, 'The Degchi Brigade', by Robert Hutchison, *The Raja of Harsil- The Legend of Fredrick 'Pahari' Wilson 2010.*
16. P. 59, Para 4: *Robert Hutchison: The Raja of Harsil- The Legend of Fredrick 'Pahari' Wilson,2010.*
17. P. 343, Para 1: Dr W Hoffmeister: *Travels in Ceylone and Continental India*
18. P. 56, 'Report of the Officials of Government of India and the People's Republic of China on the boundary Question' Part 3- 1960.
19. P. 63, Para 2: 'Report of the Officials of Government of India and the People's Republic of China on the boundary Question'- Part 3- 1960
20. P. 49, *'Notes, memoranda and letters exchanged and agreements signed between the governments of India and China- 1954-59',* White Paper by Ministry of External Affairs, Government of India.
21. P. 53-54, Col F Markham: *Shooting in the Himalayas*- 1854.
22. P. 316, Para 1-2: S J Stone : *'In and beyond the Himalayas'*-1896.
23. P. 198-206, *'Geology of the Central Himalaya'* by CL Griesbach. *Memoirs of the Geological Survey of India* - Vol 23- 1891.
24. Ibid. P. 206, by CL Griesbach
25. 'Gangotri and Leo Pargial' by Marco Pallis, *H* Vol. VI, 1934.
26. Lt JFS Ottley 'The Jadh Ganga Valley and Nela Pass', *HJ* Vol. XII, 1940.
27. J B Auden, 'A Season's work in Central Himalaya', *HJ* Vol. XII, 1940.
28. Pp 11, *'Notes, memoranda and letters exchanged and agreements signed between the governments of India and China- 1954-59',* White Paper by Ministry of External Affairs, Government of India.
29. *HJ* Vol. 68, p. 85.
30. Ibid. Auden
31. 'Jadh Ganga Valley',- *HJ* Vol. 42,1985.
32. 'Two is Company', *HJ* Vol. 47, 1989-90.
33. 'Chungsakhago',- *HJ* Vol. 66, 2010.

7

Lure of the Unknown

An exploratory climb in the East Karakoram

Divyesh Muni

As I stepped off the icy ledge and started my rappel down the steep north face of the Sakang col, I felt a surge of apprehension. This was a committing move. We were descending 500 m into the North Shukpa Kunchang glacier.

When I first sent details of our expedition to Victor Saunders I wrote 'The major concern on this route is, in the event of a mishap or ill health, getting back via the Sakang col would be a challenge and we would have to consider helicopter evacuation.' Unfortunately, this was to come true.

The lure of an unclimbed 7000'r is strong, and when you see an opportunity to attempt two, it becomes the next objective!

I had attempted Plateau Peak (7300 m) in 2009. Unfortunately, we were sent packing by the weather after reaching 6600 m and the route was exacting. The peak remained an enigma. It had been attempted by several expeditions from its western approach without any success. The route-finding itself was a challenge. It had now become one of the most sought after climbs.

Sitting unnoticed in the middle of the giant 7000'rs of the Saser group was P. 7017 m. The IMF website listed it as Tughmo Zarpo.

I toyed with the idea of attempting P. 7017 m and Plateau Peak from the eastern approach i.e. from the North Shupka Kunchang glacier. I read the articles in the *Himalayan Journals* on the two previous expeditions that had entered the valley. The North Shukpa Kunchang glacier can be approached from the Shyok valley in the east but gets blocked off during the summer months. Also, to reach the base of P. 7017 m, one would have to negotiate 32 km of the North Shukpa Kunchang glacier.

---

1. Plateau Peak was climbed in the year 2013 twice; first ascent was by a Himalayan Club expedition. For more details refer to *'Plateau Peak: The Last Unclimbed bastion of the Saser Massif'* by Pradeep Sahoo in this volume. Second ascent was made by a Ladakh Scouts team by a new route– Ed.

2. *HJ* Vol. XXXIII, p. 119

Looking for an alternative approach, it occurred to me that one could descend to the North Shukpa Kunchang glacier from the Sakang col, which I had visited during the 2009 expedition to Plateau Peak. The descent from the Sakang col would be steep and committing. But compared to the difficulty of the eastern approach, it seemed a better option.

The team consisted of Victor Saunders, Susan Jensen, Vineeta Muni, Andy Parkin and I. There was a support team of six Sherpas - Samgyal, Mingma, Ang Dorji, Chedar, Dawa and Karma, three kitchen staff - Chetup, Sonam, Kami and a runner, Ramesh. A liaison officer was deputed to accompany the team. The logistics support was provided by Rimo Expeditions.

We landed in Leh on 18 July 2013 and spent four days checking our gear, discussing plans and acclimatising. Due to the difficult terrain of the approach march, getting our loads to base camp itself was a challenge. With a night halt at Phonglas (4400 m), we were at base camp on 25 August.

We had to either find a route from the head of the Sakang glacier or make an attempt from the North Shukpa Kunchang glacier.

We rested a day at base camp settling down and unpacking the gear. We established advanced base camp at 4800 m on the central moraine of the Sakang glacier. It was a long and torturous route. Initially the route started from the lateral moraine on the true left of the glacier. We then crossed onto the central moraine to avoid crossing the subsidiary glaciers feeding into the Sakang Lungpa. A team of Kumauni porters and the Sherpas supported our movement to ABC.

On 30 July, we spent the day to do the recce of the head of the Sakang glacier. We decided to explore two options. One, to seek a route from the south face of the wall connecting Plateau Peak and Sakang Peak and the other to cross the Sakang col into the North Shukpa Kunchang glacier.

On the next day, most of the team, along with the Sherpas climbed to the Sakang col (6150 m) for an initial assessment of the route. The first sight of the descent left us speechless. 'You have a look!' was all Victor said to me. I sat silently at the col for a long time before I looked around. The Sherpas also had a grave look. This was difficult and dangerous! As I descended, I instructed the Sherpas to bring down the rope and gear that they had carried up. This was not going to work. The route looked too loose.

At ABC, we thought of alternatives and decided to have a second look at the Sakang col. On 3 August, Victor and I climbed Sakang col again. We considered a line of descent that might work. However, as we walked back to ABC, both of us had nagging doubts about the route. In the meantime, Andy and Susan explored the basin route. They were of the opinion that the route may be climbable in a quick alpine style attempt.

On 5 August, we moved to the camp below the Sakang col. The route to the camp was crevasse ridden. By mid-morning, the snow turned soft. There was uncertainty with each step… would it hold or would it collapse? One would be thigh deep in snow and sometimes find oneself dangling over a crevasse. Each one of us had to be pulled out of a crevasse at sometime or the other.

We had another look at the descent route from Sakang col. Victor, Samgyal and I traversed down some distance to the point where the route takes a vertical descent all the way to the glacier. It was a frightful experience. The route was exposed to rock fall and one could see a lot of debris strewn all along. As we stood there, we heard a couple of rocks wiz past. Without much discussion we made our decision, we would not risk descending this route. At the col, we looked further east towards Sakang Peak. About 200 m further east from where we stood, a steep ice slope fell all the way to the glacier below. There was no sign of rock fall on the slopes below. If we could find an approach to the slope, we might have a safe route down.

On the following day, Victor and I along with two Sherpas climbed up to and adjacent notch on the Sakang col ridge. For easier reference we called it ‘Andy’s col’. Susan and Andy attempted reaching a further notch directly above the proposed descent route. This we referred to as Susan’s gully.

The climb to Andy’s col was fairly straightforward. As we topped out at the col, our spirits were lifted at the sight below. YES! It looked safe and doable. A steep icy slope directly leading down from this point appeared to snake all the way down to the glacier. We walked back to camp that day very happy. Susan and Andy returned from their recce without success. So there seemed only one safe route across.

North Shukpa Kunchang glacier seen from Camp 1. (Divyesh Muni)

The next two days were spent reorganising while the Sherpas shifted loads from our initial descent route to 'Andy's' point. They also fixed about 300 m of rope on the descent and lowered most of our gear and rations half way to the glacier on the other side. Victor, Andy and Susan rested for a day at ABC before we were ready to cross over to the North Shukpa Kunchang glacier.

On 10 August, we started off with heavy loads to cross the Sakang col. Victor, Andy and Susan had carried four days of rations for themselves. Vineeta, the Sherpas and I were prepared with 12 days of provisions. We were 11 of us on the route. By the time I started down from the col, it was 11.00 a.m.

The slope immediately below was vertical for about 50 m and gradually eased to about 60 degrees. It was a step into the unknown, knowing that we could not afford to err. We were descending into the valley between two giants, Saser Kangri II and Saser Kangri III. I was quickly down the first 100 m.

As I changed anchors and prepared to move down the next rope length, Vineeta was about 50 m above me and Mingtemba Sherpa was further above. I heard the sound of an explosion. Instantly as I looked up, I saw a shower of rocks above us - I shouted but there was nothing we could do. We were restricted in our movement by the fixed rope. We dug our heads into the ice and hoped our helmets would save us. To my relief all three of us escaped unhurt. We started moving down as fast as possible. The melting ice had triggered a rock fall from the edge of the Sakang col. A massive boulder that had dislodged landed on the slope above us and splintered causing the explosion sound.

Most of the team was waiting on a rock ridge half way down the face. The route further down was on the rock band and it required careful traversing and descent. We could not afford to dislodge any rocks, since they would endanger the climbers moving below. After the rock band, we had to negotiate a snow and ice slope to reach the glacier. The Sherpas had devised a 200 m zip line to the bottom for the loads to be lowered. It took a few hours for us to gingerly climb down the slope and for all the loads to be lowered. By 4.00 p.m. we started moving down the glacier. We had to exit the narrow valley and camp in a location safe from any avalanches that may come from the Saser Kangri peaks flanking the valley.

Victor, Susan and Andy had moved ahead and scouted a safe route down the glacier to its junction with the North Shupka Kunchang glacier. By the time the rest of us reached the junction, it was nearly dark. We quickly setup camp and settled for the night. All of us were exhausted, both physically and mentally. Next morning, we had a relaxed start. The views were simply astounding. The North Shukpa Kunchang glacie, snaked down till the eye could see, flanked on both sides by high peaks. There was happiness, excitement and enthusiasm as we packed and made our way to the central glacial moraine of the North Shukpa Kunchang glacier near the southern end of P. 7017 m. It took us almost four hours to find a route across the glacier and locate a good camp site, but it was worth the effort. We called it the Chamshen camp, based on the name of P. 7017 m that we were to attempt. As we made our way across the glacier, we looked up in awe at the huge face of Saser Kangri III looming behind us. We wanted to get as far away from it as possible considering the massive hanging seracs and broken ice threatening to fall any moment.

It was now clear that P. 7017 m was not Thugmo Zarpo, which was actually an adjoining peak of lesser height. That evening, we had grand views of Saser Kangri II, Saser Kangri III, P. 7017 m, Thugmo Zarpo and scores of peaks on the North Shukpa Kunchang glacier.

On the next day, 12 August, Samgyal, Ang Dorjee and I went further up the glacier to recce a route for P. 7017 m from the north whilst Victor, Andy and Susan went east, to scout a route for an alpine attempt. Both routes looked feasible. Victor, Andy and Susan would go back across the Sakang col to replenish supplies, Vineeta and I supported by the Sherpas would attempt P. 7017 m from the north ridge.

Accordingly, we shifted to Camp 1 at 6000 m whilst Victor, Andy and Susan took a day off at the Chamshen camp. The route to Camp 1 was on the moraine for the initial section and then went onto the broken glacier. The route zig-zagged through crevasses all the way to the campsite.

As we climbed up, we could get a better view of the route towards Plateau Peak. It would be a challenge to find a safe route from Chamshen camp to the base of Plateau Peak. I spotted a possible line but we would think of that after our attempt on Chamshen.

As we settled into Camp 1, the weather packed up and it was snowing heavily by mid-afternoon. The snowfall continued through the night and the morning of 14 August did not show any signs of improvement. We decided to move back to Chamshen camp and wait out the weather. The walk back was cold and windy. With low visibility, we had to be careful whilst finding our way back through the maze of crevasses. When we got back to the camp, our British friends were preparing to leave to cross the Sakang col.

I was not happy to see them attempt the crossing in this weather. The snow accumulation, low visibility and the risk of the terrain itself were reason enough not to proceed. But they were determined and hoped that they would get across before the weather further deteriorated. They left camp and planned to spend a night at the intermediate campsite we had used on 10 August.

The snowfall continued persistently and made me very tense about our three friends somewhere on the glacier. On 15 August, Indian Independence Day was celebrated at the camp. It was snowing and we stayed put. It was late evening when I heard a voice calling out to me. It was Susan...she was in tears.

Andy Parkin being rescued after the accident. (Divyesh Muni)

On 14 August at about 10.00 p.m., the three of them had been blown over by the pre-avalanche blast of air whilst at the intermediary camp to Sakang col. The blast of the avalanche that came from Saser Kangri II was so intense that both tents were lifted off, ripping out the anchors. The tents, with their occupants inside, were thrown high, rolled over and scattered their contents as they landed. Victor and Susan had a miraculous escape. Victor injured his back, which nagged him for the rest of the expedition. Susan had some minor injuries. Andy unfortunately landed 20 m into a crevasse. Slammed against the sides of the crevasse, he had several impact injuries. He found it difficult to move.

Through the night, Victor and Susan hauled Andy out of the crevasse. They huddled into the surviving tent. Early morning brought down one more powerful blast. Although they survived this without further damage, they pumped Andy with steroids and he painfully moved two km to a safer spot on the glacier. Victor and Susan propped him in the tent and stocked some food and water with him before making their way to Chamshen camp to seek help.

The first priority now was to arrange for rescue. Our worst fears had come true. There was no possibility of Andy crossing the Sakang col on his own or even being carried across. There was no other exit point from our location. It was 32 km to the Shyok if we were to go down the North Shukpa Kunchang glacier and the river may have blocked the exit. The only way out for Andy was by air.

If we were to send a Sherpa across the Sakang col and back to Nubra, it would take several days. Our liaison officer had left the team and was

Saser Kangri I east face from Chamshen Camp. (Divyesh Muni)

Saser Kangri II north face from Chamshen Camp. (Divyesh Muni)

Summit camp at 6500 m. (Divyesh Muni)

camped with the Ladakh Scouts way below our base camp. We had to send an emergency message to arrange for Andy to be air-lifted. The only way to do this was to use the satellite phone Victor had brought for such a contingency. Although the use of satellite phones was not allowed due to security concerns, it was the only way we could arrange for Andy's rescue. We sent out the message.

In the next few hours, while we were settling Victor and Susan, Rajesh Gadgil and Rimo Expeditions had started supervising the rescue process. IMF and the insurance company were informed and a request had been sent to the Indian Air Force for air lifting Andy.

At first light on 16 August, Victor and I along with the Sherpas made our way to Andy. Within an hour and half, he was sipping hot tea and joking about his predicament - even the slightest movement was extremely painful. Victor and I broke trail through the soft snow while the six Sherpas carried Andy to Chamshen camp on a rope stretcher over soft snow with hidden crevasses. It took us several hours to get back to camp.

In the meantime, we had provided our exact location and the Air Force team charted the course for the helicopters to reach us. It had to be a long circuitous route along the Shyok and up the North Shukpa Kunchang glacier. Due to the low visibility, the choppers could not fly. The pilots were on standby, ready to fly as soon as the weather provided a window.

On 17 August, we waited eagerly for choppers to arrive. We had prepared a helipad by stamping down the fresh snow and placing markers and a flag to indicate the wind direction.

We had resigned to wait another day for the rescue. Just then, at about 5.00 p.m. the distant sound of rotors caught our attention and brought cheers all around. As we scrambled out of our tents, pulling on boots and jackets, the helicopters came to view flying just above the glacier keeping below the low clouds. Andy was literally hauled out of his tent and carried by Ang Dorjee on his back. The helicopters flew Andy off to the hospital in Leh after an overnight halt en route.

Andy was later diagnosed with a fracture of the sacral bone due to which he was in intense pain and could not move. Hats off to all those involved in the rescue, specially the pilots, whose skilfull flying in such difficult conditions got Andy out.

We breathed a sigh of relief, knowing that Andy was in safe hands. The weather also appeared to settle. With a few more days to spare, we had to decide our further course of action. We were now running short of rations. Victor and Susan were also in a dilemma, whether to cross the Sakang col or to join us for an attempt on Chamshen. They had no food supplies with them and would be dependent on us. We gladly agreed to share our rations should they choose to stay on and join us for the attempt on Chamshen.

Vineeta meticulously took stock of the rations and budgeted the same for the remaining days. We were now on survival rations. We waited a day for the weather to settle and shifted on 19 August to Camp 1 at 6000 m for the first and last shot at the summit. On 20 August we shifted to Camp 2 at 6500 m. The route varied from 400 to 500 and we had to negotiate a few large crevasses. 600 m of rope was fixed on this section. We had now run out of fixed rope. Not knowing the difficulties of the terrain to the summit, we retrieved the top 200 m of rope to carry with us for the summit attempt.

Climbers on the summit ridge of Chamshen. (Divyesh Muni)

We had to negotiate about 1500 m of the west ridge to gain 520 m of altitude. On 21 August, we started at the break of dawn and slowly made our way on the summit ridge. The route was fairly simple with a few sections of steep snow and one stretch of steep loose rock. We were fortunate to find easy passage past two rock gendarmes en route. By 11.00 a.m. the last of us reached the summit to enjoy a grand panorama of the East Karakoram peaks. We spent a long time celebrating, taking pictures and just soaking in the grand views around.

Back to Chamshen camp on 22 August, we spent the next day resting and preparing to cross the Sakang col. The narrow valley leading to the col was now referred to as 'The Valley of Death'. We decided to cross in one shot, leaving the Chamshen base on the midnight of 23 August.

We made quick progress in the moonlit night. As we passed the accident site, Victor could retrieve some of the lost equipment. By early morning we were at the base of the col. The terrain had changed dramatically. The 500 m climb to the top of the col was one of the most frightening experiences due to the constant threat of rock fall, the steep ice and the enormous loads. The lower fixed ropes were completely swept off by avalanches. Huge avalanche debris at the base of the col looked ominous. We climbed as fast as we could to reduce our exposure. But the prolonged exposure over the last several days to altitude and the lack of nutrition was taking its toll. Our movement on the slope was

Summit ridge of Chamshen. (Divyesh Muni)

very slow. The fixed rope on the rock band was damaged in several places. Crossing the last 200 m to the col on the near vertical ice with heavy loads was the most difficult climb of the expedition. Exhausted and dehydrated, most of us were hauled across the last overhanging section of ice by a team of Sherpas with an additional rope. We were more elated getting across the col, than we were on reaching the summit of Chamshen!

Over the next few days, we wound up all camps. At base camp, we were greeted with the news of successful ascents of Plateau Peak by members of The Himalayan Club Kolkata Section and the team of Ladakh Scouts by different routes. At the roadhead, however, we were greeted by the Police, who were inquiring about our use of the satellite phone. Victor was arrested and let off with a minor penalty and confiscation of the phone instrument. The Magistrate at the Leh court was considerate since we had used the phone only for an emergency. We hope that some system can be worked in the future where legitimate use of satellite phones is permitted by the Government of India and security agencies.

We had an immensely satisfying expedition. Andy was safe and recovering from his injuries. The team had achieved the first ascent of P. 7017 m, which we called Chamshen and we had made the first and probably the last crossing of the Sakang col.

Summary

On 21 August 2013, an Indian-British joint expedition achieved first ascent of P. 7017 m (named Chamshen) in the East Karakoram, after making the first crossing of the Sakang col (c. 6150 m) linking Sakang glacier and North Shukpa Kunchang glacier. On 14 August, Andy Parkin met with an accident due to an avalanche blast and was seriously injured. He was saved by helicopter pilots from the Indian Air Force, who were contacted through the illegal use of a satellite phone by the team.

8

Plateau Peak -The last unclimbed bastion of the Saser Massif

Pradeep C. Sahoo

> Faith is taking the first step, even if you have not seen the whole staircase.
>
> - Martin Luther King Jr.

It snowed heavily through the night. I remained huddled with my wife Chetna, inside a small arctic tent at Camp 1 on the South Phukpoche glacier, deep in the eastern Karakoram. As I opened my tent's flap and peered out, I was welcomed by the sight of a thick white blanket of snow covering the lateral moraines of the glacier. A heavily overcast sky, poor distant visibility, absence of wind and incessant snow fall always portends the appearance of a spell of poor climbing weather. A clime in which one can do little but to huddle up inside warm sleeping bags and wait for the sky to clear. Fortunately, we were exhausted and welcomed this opportunity to rest after the last three days of sustained climbing, including twenty one hours from Camp 3 to the top of the Plateau Peak and back.

The joy was slowly sinking in – the Kolkata section of the Himalayan Club had ascended the last unclimbed bastion of the Saser peaks via the sheer and broken west ridge of the Plateau Peak (7300 m).

The Saser mountains have five peaks above 7000 m. All these peaks offer significant technical mountaineering challenges on their steep ridges and faces. Saser Kangri I (7676 m), Saser Kangri II main (7518 m), Saser Kangri III (7495 m) and Saser Kangri IV (7410 m) had been ascended between 1976 and 2011. Though several attempts were made on Saser Kangri V or the Plateau Peak, both on the western approach from the South Phukpoche glacier as well as the eastern approach from the Sakang glacier, success eluded all previous attempts.

We had had a close look at Plateau Peak during our ascent of Saser Kangri IV in 2011[1] . The face of the mountain seemed to be guarded by dangerously balanced seracs from the summit plateau. The only

1. *HJ* Vol. 67, p. 272

Camp 1 at South Phukpoche glacier and Plateau Peak. (Plateau Peak Expedition 2013)

feasible route appeared to be its western ridge, but the sheer rock wall of the western ridge seemed a serious bottleneck. After our successful ascent on Mamostong Kangri (7516 m) in 2010[2], Saser Kangri IV (7416 m) in 2011 and Jongsong East (7447 m) in 2012[3], our climbing team appeared confident in tackling the intricacies of the west ridge. A team of nine climbers under the leadership of Debraj Dutta set about planning the expedition. The veteran Phurba Sherpa, who has been a permanent member of our team since last four expeditions, led a seven member Sherpa team from Darjeeling with three kitchen staff. They joined us in New Delhi.

The entire team congregated at Leh on 4 July 2013 reaching the roadhead at Phukpoche on 6 July. This is a five hour drive from Leh across the Khardung la following Nubra river. 7 and 8 July were spent acclimatising, arranging locals to ferry our load to base camp and fixing ropes en route to the base camp for the safety of the climbers as well as the local people carrying our load along the deep gorges of Phukpoche nala. Getting adequate porters for large expeditions in the Nubra valley is a perennial problem, but our contacts at Phukpoche came handy,

2. *HJ* Vol. 67, p. 145

3. *HJ* Vol. 68, p. 65

particularly Wangchuk, who also offered the extension of his house for our large team to stay.

The advance team led by Pasang and Migma left with the first batch of load ferrying porters on 09 July. Over the next five days, some ninety loads of twenty kg each were moved to the base camp at the confluence of the South Phukpoche and the North Phukpoche glaciers, at an altitude of *c.* 4700 m. By 14 July, the entire team was at BC. In the meanwhile, our advance team had already commenced dumping loads at our proposed Camp 1 site on the South Phukpoche glacier.

By 15 July, our Camp 1 had enough supplies and climbing gear. 16 July was a rest day and we performed our customary base camp Puja with Hindu and Buddhist rituals. On 17 July, an advance team consisting of Chetna, Pradeep, Phurba, Pasang, Migma, Dawa and our cook Indra occupied Camp 1 (*c.* 5400 m), while the others ferried loads. Camp 1 was in the middle of the South Phukpoche glacier, close to its true right bank on a fairly flat portion. This had served as Camp 1for our 2011 expedition to Saser Kangri IV.

The South Phukpoche glacier drains out from the accumulation zone of Saser IV, Saser I and the Plateau Peak. Two parallel ridges descend from Saser IV, The west ridge guards the north side, and the south west ridge of Plateau Peak guards the south side of South Phukpoche glacier. Saser IV, Saser I and Plateau Peak create the eastern boundary, giving a wonderful amphitheatre like view of peaks and ridges. However, they tend to block the clouds moving in from the west across the Nubra valley hence influencing local weather conditions. The west ridge of the Plateau Peak falls sharply from the summit plateau - this was clearly visible from our camp location. This approach was tried several times in the past by climbers from the Indian armed forces.

The BSF team in 2009 had successfully tackled the lower part of the west ridge up to a height of c .6400 m but was stopped by a rock tower. They also faced a sustained period of bad weather leading to their expedition being aborted. From our Camp 1 location, we did a thorough investigation of the route using our tele-lenses. Our team's experience on the south ridge of Saser IV and on the south face of Jongsong, where we had handled highly technical and long stretches of rock climbing had given us experience and courage to handle this rock pinnacle. We also worked on a backup plan, which was to circumvent the rock

Climbing the rock band. (Plateau Peak Expedition 2013)

pinnacle. This climb would require a dangerous traverse on the highly exposed south face of the Plateau Peak that overlooked a sharp fall of around 600 m to the South Phukpoche glacier and was also exposed to constant threat from huge seracs hanging precariously from the summit plateau. A short traverse of about 200 m or so would however get us to the safety of the western ridge. The southwest ridge, as viewed from our base camp location had offered an alternate approach, but from our Camp 1 location, we observed that there was no access to the southwest ridge from the South Phukpoche glacier. During the course of the expedition we found a team from the Ladakh Scouts was attempting this route, but they had approached the mountain from the other side (Sakang glacier).[4]

On 18 July, our advance team led by Phurba opened the route to Camp 2 (*c.* 6000 m) on a fairly broad shoulder on the lower part of the west ridge. The route to Camp 2 from Camp 1, involves crossing the entire breadth of the South Phukpoche glacier diagonally from our Camp 1

4. The Ladakh Scouts team were successful in reaching the summit of Plateau Peak within few days from the Himalayan Club team's ascent. Unfortunately, no further details could be obtained of the ascent from the Ladakh Scouts team. – Ed.

location to the base of the ridge and then climbing about 200 m of a rocky slope of 40 to 50 degrees and then an ice slope of 30 to 40 degrees. The glacier traverse did not offer much difficulty as it was either moraine covered or compacted with hard ice without many crevasses. The shoulder at an approximate height of *c.* 6000 m had a snow field about the size of a football field - an excellent site for Camp 2. We camped close to the rock buttress.

On 19 while Phurba, Mingma, Mingma Thendup and Lakpa occupied Camp 2, Lakpa, Gyalzen, Dawa, Passang and I ferried load to C 2. The same day Debraj, Prashant, Subrato, Ganesh and Adrito occupied Camp 1. Biplab, Parag and Karma remained in base camp to coordinate the supply lines. We spend 20 July exploring the glacier around Camp 1, while the advance team with Phurba worked on the west ridge above Camp 2. The initial stretch above Camp 2 involves climbing the rock buttress which then turns into a sharp ridge of broken rock and ice. It was a tricky traverse that entailed walking very gingerly on the dodgy ridge and sometimes making a sideward traverse on the steep slopes that fell precipitously to the South Phukpoche glacier below. The entire route beyond the top of the buttress required to be secured with fixed lines. We saw two old dilapidated lines from the previous expeditions - an eight mm nylon rope and a 10 mm plastic rope. That day the advance

Negotiating lower portion of the west ridge after Camp 2. (Plateau Peak Expedition 2013)

team reached the bottom of the rock pinnacle, which was the high point reached by the BSF team in 2009. They located a possible camp site at the bottom of the rock pinnacle, which also seemed to be the camp site for the previous expedition. A total length of 1200 m rope (four coils of eight mm plastic rope) was fixed to secure the route.

On 20 and 21 July, the advance team worked to explore the route across the rock pinnacle, while the others made a load ferry to Camp 2. The rock pinnacle was found to be too difficult to establish a feasible route upwards, so Phurba and his team, worked on our plan B - to find a route around the rock pinnacle. Starting 5.00 a.m. in the morning on 21 July, they moved quickly to the bottom of the rock pinnacle and then concentrated on circumventing the steep pinnacle. By late in the afternoon, they were successful in fixing two coils of rope (600 m), one on the traverse on the south face and the other on a slope of blue ice connecting to the bottom of the upper rock band. We were now successful in cracking a route across the rock pinnacle. This was a hazardous route and we planned to cross that portion in the night as that would minimise our exposure to rock fall and ice avalanche. The advance team of four climbers had been working continuously for last five days. On 21 evening they came down to Camp 1 for a well earned rest, while the rest of the team stocked up Camp 2 for the remaining days of the expedition.

Our big team together enjoyed a rest day on 22 July in Camp 1. We worked out the logistics for the remaining portion of the climb. The advance team would continue working on the remaining portion of the rock wall on the upper portion of the west ridge. The other teams would occupy Camp 2 a day later and then we would form the summit team that would occupy the summit camp / Camp 3 (*c.* 6400 m) below the bottom of the rock tower. Our leader Debraj, detailed out the plan to all members on 22 evening and we retired early keeping in mind the hard day ahead.

The morning was not the best as we woke to moderate snowfall. More worrying was the cloud condition both on the summit of the Saser massif and the dark clouds coming up from the Nubra valley through the Phukpoche nala. An early start was not possible as we wanted to gauge the mood of the weather and take a call as the morning progressed. To our dismay, the adverse weather showed no sign of relenting. By evening even though the snowfall had stopped, the overcast condition persisted. The full moon period had just passed by. The Sherpa members usually have a natural premonitory ability about the high altitude weather trends and they believe that if the weather turns foul around full moon, it will remain so for at least four / five days. We prayed for the weather to improve but 24 and 25 July also exhibited similar weather conditions. We were worried about the snow conditions in the higher realms of the mountain. On 25 July we sent three of our Sherpa members to base camp, braving bad weather, to bring up fresh supplies to Camp 1. The morning of 26 July saw another day of heavy snowfall. We were now into the fifth day of inactivity at camp and this made us feel very restless. That evening, the weather finally showed some improvement and later we saw the stars glittering and the dark clouds slowly moved away.

As per our earlier plan, the advance team consisting Phurba, Mingma, Thendup Lakpa, Dawa, Gyalzen, Chetna and myself occupied Camp 2 the next day in excellent weather. Camp 2 offered an unhindered view of western Karakoram mountains including the mighty Broad Peak and Gasherbrum. For the first time we saw the Ladakh Scout team on the lower portion of the southwest ridge through our telephoto lenses.

On 28 July, at 4.30 a.m., Phurba, Migma and Lakhpa commenced the opening of the remaining portion of the upper rock band in order to find a safe passage through the hanging seracs to the lower plateau of the summit. Later in the morning Thendup, Gyalzen and Dawa carried loads to C 3 at the base of the rock pinnacle.

Phurba's team had a tough job retrieving the ropes buried under several inches of snow over the past five days. By 10.00 a.m. after toiling continuously for over six hours they reached their highest point on the lower rock band. By 2.00 p.m., Debraj, Subrato, Prasant, Ganesh and Adrito reached Camp 2. A little later the team that had gone to drop load at Camp 3 also returned to Camp 2. All the members were in excellent condition except Chetna, who had a nagging cough that had been troubling her from Camp 1. It was a very bright afternoon. All eyes were set on the trio making their way up the tricky upper rock band. By 5.00 p.m., we saw them between the top of the rock wall treading slowly on an ice slope that lay between the final rock tower and the seracs of the lower plateau. As they had only two and half hours of day light left to descend safely, we persuaded them over the walkie-talkie to come down as the most difficult part of the west ridge was behind them. But Phurba wanted to make sure that he had a view of the western plateau before they returned. By 6.00 p.m., we saw them on the top of the highest dome on the lower plateau seen from our camp. Then the trio slowly started descending on the fixed lines. It took them nearly four hours to reach Camp 2. But they had completed an excellent job on the wall which was the main hurdle on the west ridge.

The following morning, we were to occupy Camp 3 for our final summit attempt that night. The advance team was not in a good shape and the second team that had occupied C 2 also needed a day to acclimatise, before making the final push. Though the weather was good, we decided to take a rest day before occupying the summit camp. Chetna was still not fully recovered from her cough, which started worsening. She decided against going to the summit camp and planned to return to Camp 1. Today, we could not see the Ladakh Scout climbers who probably were taking a day off too!

Camp 3. (Plateau Peak Expedition 2013)

On 30 July, Debraj, Subrato, Prashant, Adrito, Ganesh and I and seven of our Sherpa members left Camp 2 leisurely at 8.00 a.m. as the climb to C 3 was just over four hours. The plan was to reach C 3 before one in the afternoon and after a good rest to resume the final climb at mid night.

The initial climb was through a large heap of loosely stacked boulders above Camp 2, where we moved without our crampons. The ridge subsequently turns into a knife edge, which

Near the western end of the summit plateau. (Plateau Peak Expedition 2013)

required some tight rope walking on the knife edge and some horizontal traverses on the steep slopes. After gaining about 300 m of vertical height, we descended about 30 m on to a depression on the ridge which was rising sharply to the base of the rock pinnacle. The rocky outcrops on the ridge were good for fixing rock pitons. By one in the afternoon, all of us were at Camp 3, and we pitched three tents for the twelve of us. In the evening, the panorama of numerous peak of western Karakoram, glowing in the setting sun was breathtaking. The summit team consisted of Debraj, Subrata, Prashant and I along with Phurba, Migma, Thendup, Lakpa and Dawa. Gyalzen and Lakhpa Sr. stayed behind with Ganesh and Adrito as a support to the main team. I could hardly manage any sleep. At 11.00 p.m., Lakhpa Sr. brewed some tea which we had with corn flakes. By mid night, we set out for the traverse around the rock pinnacle. It involved navigating precariously through a few seracs and then crossing a bergshrund. We walked in the pool of light from our headlamps, oblivious of the lurking dangers from the seracs or the dangerously exposed south face of Plateau Peak. We climbed about 200 m of blue ice at a gradient of 50 to 60 degree, before we gained access to the rock wall again.

There was no wind; the sky was clear with the glittering stars and the half moon. We inched ahead on the fixed line mechanically. The lower

Saser Kangri II seen from the summit. (Plateau Peak Expedition 2013)

Saser Kangri III seen from the summit. (Plateau Peak Expedition 2013)

portion was a mixed terrain of rock and ice, giving us enough purchase for our crampons. As we moved towards the upper section of the rock wall, it was a tricky maze of chimneys, narrow rock gullies covered with a thin layer of ice, making our climb extremely tricky. By day break we were in a 15 m high chimney, just wide enough to climb in a bridging position. After the chimney was a vertical wall of five m. Once we were above the rock wall, we got into a narrow slope of ice at 40 to 50 degrees and 10 m wide between the top of the rock wall and the hanging seracs of the west end of the plateau. This slope finally connects to another large serac by a narrow snow bridge. We cautiously made our way through to the top of the serac and then passed another large serac about the size of a five story building on our left as we climbed up to the western end of the summit plateau. By 9.30 a.m. the first five members reached this balcony, the others caught up in half an hour's time.

After the excruciating climb of ten hours, we rested for 20 minutes and started our long traverse across the summit plateau. A vast rolling snowfield unfolded in front of us. It appeared as if we were on a polar landscape. After covering about half of the plateau, we could see Saser Kangri IV and Saser Kangri I summits emerging towards the northeastern direction. Saser Kangri II and Saser Kangri III were still out of sight. The snow condition was perfect; it was hardly a couple of inches thick. The weather God was kindly disposed towards us and offered us a perfect summit day. After trudging along the broad ridge for about an hour we perused Saser Kangri II on our right. Saser Kangri III was still not visible due to the presence of humps in front of us towards the northeast which constitute the highest location on the mountain. The closest hump was approximately 200 m ahead of us. By 1.30 p.m. we reached the highest point. When we reached the top of the hump, both Saser Kangri II and III became visible. We took extensive photographs in celebration of our success on the mountain. The nine summiteers were Debraj Dutta, Subrata De, Prasanta Gorai, Phurba Sherpa, Lakpa Sherpa (Sr), Mingma Sherpa (Sr), Dawa Sherpa, Lakpa Sherpa (Jr) and I. From this point we could see all four peaks of the Saser group, i.e. starting from the north - Saser Kangri IV, Saser Kangri I to its right, Saser Kangri III and Saser Kangri II further to the right. To the left of Saser Kangri IV, we could see Mamostong Kangri, Rimo group, Teram Kangri and eight thousanders like K2, Gasherbrum, Broad Peak and Nanga Parbat. Apart from these many lofty peaks were visible towards the northwest, which we photographed. The ridge of the Plateau Peak spans more than 1200 m with small humps dotted along its length. Our expedition team covered a large stretch of the Plateau ridge and headed towards northeast as indicated in the Russian contour map.

Unlike a conventional mountain that has a triangular appearance, here on the Plateau Peak, true to its name, we traversed for four hours gaining a height of about 200 m from the western end of the plateau towards the eastern end of the plateau. The excellent technical work undertaken by Phurba and his Sherpa team on the rocks of the dangerous west ridge was one of the finest efforts on sheer rock at such high altitude. Our maiden success on Plateau peak was principally facilitated by their climbing skill coupled with excellent teamwork. This effort led to a large team of nine climbers succeeding on a virgin peak that had eluded many brave attempts in the past.

It was doubly tricky during the descent. The radiant sun during the day had melted the ice around the ice screws. Migma moved in the front in order secure them to ensure the safe descent of the team. By 7.30 p.m., when darkness set in, we had descended the rock wall and had reached the upper slope of the rock pinnacle. It was a frightening feeling to see the dangerous slopes that we had negotiated in the dead of the night. We switched on our head lamps again and continued our slog. By about 9.00 p.m., we discerned the faint lights of Camp 3, where four of our members were stationed. It was an emotional welcome and they assisted us in untying our crampons and in removing heavy boots and seat harnesses.

On 1 August by late afternoon we reached Camp 1 to a joyous welcome.

The team finally came down to Phukpoche on 7 August and reached Kolkata on 12 August to a warm reception by the media and the mountaineering fraternity in Kolkata.

Team members:

Debraj Dutta (leader) , Subrata De (Deputy Leader), Pradeep C Sahoo, Prasanta Gorai, Phurba Sherpa, Mingma Sherpa (Jr), Migma Sherpa (Sr) , Lakpa Tenzing Sherpa , Dawa Sherpa (Asst Cook) , Lakpa Norbu Sherpa,Chetna Sahoo, Parag Kr Mitra, Biplab Banerjee, Adrito Paul, Ganesh Saha, Pasang Gyalzen, Karma Thimle Sherpa (Asst Cook), and Gyalzen Sherpa and Indra Rai (Cook).

Author's Note:

Our team did not have access to the Survey of India contour map of this area and had to refer to the Russian contour map which mentions the highest point as 7288 m which nearly matching the height recorded in the list published by Indian Mountaineering Foundation – 7287 m. The Spanish map, on the other hand, shows the highest point of the Plateau Peak to be 7300 m. Thus we can see a difference of heights of the same peak as mentioned in the IMF list, the Russian map and the Spanish map. The IMF may like to reconcile this difference[5].

Summary

First ascent of Plateau Peak (7300 m) achieved by a large team from the Kolkata Section of the Himalayan Club. They approached the mountain from the western side via Phukpoche glacier. This peak was attempted many times in the past without success.

5. The latest available Survey of India maps show the height as 7300 m – Ed.

9

Climbing in the Rongdo Valley

Joie Seagram

We were a modest group of four climbers: three Canadians from Golden British Columbia, (Dr. Jeff Dolinsky and his wife Joan and me) and one American from Bishop, CA (Andy Selters). We were accompanied by climbers Kunzang Sherpa (Sirdar), Arvind Raman (LO), two climbing staff Da Nuru Sherpa (Dawa) and Nangang Bhote, and three staff: Raj Kumar Rai, Mu Gombu Rai and Chamu Singh.

Our climbing objectives were located at the head of a beautiful and remote valley called Rongdo in the Nubra area of Ladakh, lying between the east and west arms of the Shyok river and north of the Indus. Tom Longstaff travelled close to what is Rongdo village today, while exploring the Siachen glacier and Saltoro areas in 1909. It gave me quite a thrill to look across the lovely, wide meandering Shyok and imagine Longstaff moving slowly up the opposite bank on his way north – to be so close to his footsteps!

Unclimbed peaks east of Rongdo I. (Joie Seagram)

I initially called the four peaks (all just over 6000 m) forming the natural cirque, Rongdo I to IV, but we have since applied local names indicative of Tibetan culture which forms the foundation of Ladakhi society.

Upon arrival in Leh (3500 m) on 31 July 2012, over the next few days we went through the usual acclimatisation routines and busied ourselves with last minute preparations. Despite following this routine, Jeff developed a serious case of pulmonary edema requiring him to spend the better part of three days with an oxygen tank. Jeff and Joan remained in Leh an extra five days for acclimatisation.

While Jeff and Joan remained in Leh, Andy, I and our support staff proceeded by jeep to cross the Khardung la and made our way over the mighty Shyok river to Rongdo village (3000 m). All was well in charming Rongdo (a dozen dwellings) except for one minor problem – a shortage of horses and attendant pony-men to transport our loads up the valley! Most villagers and valley folk were engaged with a visit by the Dalai Lama in Leh, while the remaining friendly locals appeared quite perplexed by our presence indeed, being unaccustomed to seeing tourists in Rongdo. A few quick calls to Leh and six horses were arranged to be sent over the Kardung la in the morning – I was horrified! How could any beast would survive such a long and circuitous mountain journey? Nonetheless, six horses arrived next morning, looking healthier than most of us. These six, in addition to a few scattered local horses and ponies along with the odd donkey, made our domestic hauling team complete.

On 05 August we slowly began our walk up Rongdo valley heading northeast, following the south side of Rongdo river. It was truly a joyous experience. Over five days we wandered past ancient stone dwellings and eroding Buddhist chortens and stupas, well-tended barley fields alongside rich potato patches, finally culminating in a visit to the well-preserved Rongdo gompa – a small Tibetan Buddhist monastery located in this very remote valley for well over 100 years. Most astonishing to witness as we wandered along, were the gigantic (hundreds of metres high) solid looking walls jutting up both sides of the valley – a rock climber's paradise (indelibly added to 'the list'). In addition, the understated beauty of dry valley hillsides dotted with sprays of green, and scattered with healthy dzo's unfolded before us, while steep snow clad peaks towered in the backdrop.

We continued to be intrigued and deeply touched by our encounters with the local herdsmen and farmers of Rongdo valley. They in turn seemed to share our curiosity and mutual comfort with strangers. Andy's advanced linguistic skills proved to be invaluable in helping everybody communicate and enjoy each other, for we were as foreign to the Ladakhis as they were to us. Rongdo has seen virtually no tourism, save one Indian-American climbing team in 2005[1] which descended the valley in two days, from the Satti area due north of Rongdo valley.

Upon reaching the head of the valley we pushed onward to establish base camp (4802 m) on 10 August, on a comfortable green area with handy water and ample feed for our hard-working horses. Over the next few days we explored higher, trending southeasterly toward the cirque in search of a suitable advanced base camp. Our focal point was a large glacial lake draining the unnamed glacier around which Rongdo peaks I to IV are situated. I was relieved when we eventually found this lake (just where the map indicated), as sly comments were beginning to surface like 'there's a lake' with group members smiling and pointing to tiny tarns! With my credibility restored, we all looked down from huge boulders, in silent askance at the vast lead-grey liquid expanse – all feeling the same awe and incredulity – nobody had stood by this lake before (to the best of our knowledge). Talk about bonding!

By late afternoon on 14 August we happened upon a perfect green patch just below our main climbing objective Ngapo Kangri or Rongdo I (6350 m), which we designated ABC (5181 m). From this glorious grass-covered camp, suitable for casual football and frisbee, we could see the unnamed glacier and cirque with the Rongdo peaks (Rongdo I to IV moving clockwise). After briefly resting and relishing in our find, we returned to BC to behold yet more good news. Upon arrival we could see small specks in the distance moving slowly up the trail from the valley below. It could only be Jeff and Joan making their debut! How exciting for everybody, as we had all quietly begun to see ourselves as a much smaller group. We had no way of knowing whether they would make the journey or not. Now, we were complete.

After several reconnaissance trips around Ngapo Kangri, and up the main glacier to view the cirque of Rongdo peaks, on 18 August Andy,

1. *HJ* Vol. 62, *AAJ* Vol. 48

Jeff and Joan, Kunzang, Dawa, Nangang, Arvind and Gombu climbed the west sub-summit (6000 m) of Balden Lhamo (Rongdo III) via the col between Rongdo III and Rongdo IV, then up the southwest ridge (10 hours return). I took a rest day after recceing the northwesterly slopes of Ngapo over the two previous days to 6000 m. Climbers returned tired from Balden Lhamo and after a tasty dinner retired early.

19 August brought a minor set-back as it became clear throughout the morning that Andy was unfortunately suffering from a minor case of cerebral edema, probably due to dehydration and overexertion the day before. He had spent an especially hard day making steps in deep snow and staying in front most of the day above the glacier. After careful observation and discussion through the day, at 4.00 p.m. Andy finally agreed that he was indeed experiencing cerebral edema, and he and I descended with Raj and Nangang to BC. After two hours at the lower elevation and oxygen at 4L per min, Andy's appetite returned and he felt 100% better! Patience, communication and determination to be preventative, had indeed helped to avoid a more serious problem Andy has wide Himalayan experience and this can make it difficult for anyone to understand why, at a relatively lower elevation one should get sick.

On 22 August Jeff and Joan, Kunzang, Dawa, Nangang, and Arvind climbed Chamba (6170 m) or Rongdo II via the southeast glacier (10

Nearing the summit of P. 6160 m (Gazgazri). (Andy Selters)

P. 6160 m showing west and south faces. (Joie Seagram)

hours return). On 23 August Nangang and I climbed the upper west rock ridge of Ngapo Kangri to 60 odd m below the corniced summit (12 hours return). At the same time Andy and Arvind ascended Ngapo Kangri via the southerly slopes; first on talus and rock ledges, then up the avalanche prone southeast aspect to easier angled summit slopes, finally gaining the corniced summit (14 hours return).

On 27 August Kunzang, Dawa, Arvind and I headed up the main glacier to attempt the true summit of Balden Lhamo (6120 m); high camp on the glacier at 5690 m. An early departure next morning got us up the icy headwall to the col (6060 m) by 09.00 a.m., where the steep west ridge curves south to become the long exposed summit ridge. We deliberated in the chilly col and decided finally, disappointed, to retreat in the face of poor weather and more technical ice than we were prepared to climb. On 29 August Andy and Nangang left a high camp at 5181 m, several km northeast of our ABC, to climb Gazgazri or P. 6160 m. They ascended the southwest ridge, then traversed onto the icy south face, after which several short pitches the last being up to 70 degree hard ice, led them to a ledge at the edge of the summit dome, whence they ascended to the highest point. Descent was via the west face (14 hours return to BC).

On 30 August we awoke to chilly temperatures, steel grey skies and several cms of snow. It was regrettably; time to leave our magic hidden playground. By 31 August we were all ensconced at BC looking forward to the evening's Full Moon celebration. We enjoyed a close and memorable time singing freely together around a huge dung fire – a well-harmonised international climbing group! Next day we began our descent in the long enchanting valley, taking two days to arrive back at Rongdo village.

As we savoured our journey and modest climbing achievements together over our last delectable dinner in Rongdo village, we had a collective sense of pure amazement. That we had had the incredible fortune to spend a month in an untouched valley, climbing unclimbed peaks, and reliving our childhoods – such was our freedom of the hills…

Notes

1. 'Ngapo' (Rongdo I) is Ladakhi for (male) Blue Sheep.
2. 'Chamba' (Rongdo II) means future Buddha.
3. 'Balden Lhamo' (Rongdo III) is the name of a female goddess (Rongdo Gompa).
4. 'Gazgazri' (Peak 6160 m) is the local Ladakhi spelling for Lammergeier (Bearded Vulture).

Summary

Exploration and climbs in the Rongdo valley of Ladakh by a Canadian-American team during August 2012.

10

Rongdo Valley Exploration

Gp. Capt. V. K. Sashindran

A team of 12 mountaineers from the Indian Air Force explored the little known Rongdo valley in May – June 2013. This valley has seen only two forays in the past. The team spent 35 days in the valley, explored the main and major side valleys and, climbed seven virgin peaks and attempted one more in the area.

Rongdo Lungpa is a small stream originating in the glacial mass of East Karakoram north of the Shyok river in the Nubra region. The Rongdo village is located at the mouth of the stream and is 35 km from the bridge across the Shyok at Timur and 52 km from Hunder. Satti, a larger village with a primary health care centre, is its nearest neighbour and is located 20 km away to its west and is connected to it by a dirt track.

For the sake of easy description, the Rongdo valley can be divided into two parts: the lower and upper Rongdo valleys. The lower Rongdo valley extends from the junction where the Rongdo Lungpa meets Shyok river just below the location of the Hot Springs. The former also corresponds to the location of the Rongdo village and the latter to the place where Tara Tokpo from the south and Koyak Lungpa from the

Base camp at the mouth of Tiburchan valley. (IAF 2013 Expedition)

north join a stream Sakskyulas Tokpo coming from northeast to form Rongdo Lungpa. This marks the general upper limit of the summer grazing area of the Rongdo villages. Upper Rongdo comprises of the Sagtogpa glacier from which two streams originate and join to form Sakskyulas. There are eight side valleys on the true left side and four on the right. The valleys are referred to by the side on which they are located and numerically starting from the head of the valley. In upper Rongdo, there are three side valleys. All others are in lower Rongdo.

Immediately beyond Rongdo village, the valley narrows due to a large rockfall on the true right side. The trail is about one m wide in lower Rongdo and used for regular movement of people, cattle and flocks of sheep moving up and down the valley. 500 m beyond the village, the trail starts climbing and snakes up the right side of the valley. It ascends steeply for about 400 m and then descends more gently to a small clearing with many trees and brooks running through it called Changma (*c.* 3830 m). From Changma the trail goes up and down scree on the left side of the valley till you reach Thipti 20 minutes away. The trail stays to the right of a clump of willow trees with some pasture land enclosed by a fence of thorny branches. A small adobe hut perched on a huge rock within the enclosure cannot be missed. Doksa *c.* 4030 m is an hour's walk from Thipti. Here the valley widens considerably. Rongdo Lungpa runs along the right side of the valley and to its left is a large meadow which has now been brought under plough and serves as arable land for Rongdo villagers. Numerous rivulets coming down from glacial melts high up on the north of the valley serve as convenient irrigation channels. Beyond Doksa, the valley remains wide all the way up to the confluence of the Rongdo Lungpa with the stream coming down from the second valley on the left. This valley is called Chudon nangma which can be translated as 'hot water' valley. This refers to the hot springs located to the east of the confluence of the two streams. Nangma is the Ladakhi word for a big valley. This valley is also known as Chhurn nangma. The expansive meadows here make good grazing and camping grounds.

On the way up the main valley, we followed the Sakskyulas Tokpo further up to where a tributary from the first eastern valley on the right (R1) joins it. We crossed this stream and set up our first base camp- BC 1 (*c.* 4790 m) on a meadow to the left of this tributary. This site is a 45 minute walk from the hot springs. A broad peak is visible to the true right of the main valley just as you approach the base camp. The peak, like a truncated pyramid, was our first objective. A reconnaissance trek up the main valley showed that this peak is part of a massif that extends along the right side of the upper Rongdo valley all the way up to the head of the valley. The steep face, ice overhangs and cornices that were visible on the western aspect of the mountain made it clear that it could not be climbed from the main valley side (i.e., east face). Moreover the scree and large boulders at its base made going up the valley difficult. Two streams come down from the Sagtogpa glacier at the head of the main valley. The right stream begins at the foot of the Ngapo (Rongdo I), a 6350 m peak and flows below the terminal moraine blocking the entrance to the first side valley on the left. This valley was explored by a Canadian expedition in 2012[1]. This is a broad valley and has many peaks above 6000 m hedging it. There is a cirque at the head of the valley now named Canadian cirque. In mid-May, upper Rongdo was still draped in snow. The snow over the boulders made the going difficult and we had to walk along the streams to reach the head of the valley. After the initial reconnaissance, we turned our attention to Theorchung lungpa. We climbed up the ridge on the true left of the valley and followed the ridgeline. There are many cairns on this ridge erected by shepherds and the valley is thus called Theorchung lungpa (Tiburchan valley in Survey of India maps). 'Theor' is the Ladakhi word for cairn and lungpa is the word for a narrow valley / river. It takes three hours to reach the head of the valley where we established our advance base camp (ABC 1) (*c.* 5305 m). Half a km ahead to the northwest was a cirque which separates this valley from a glacier referred to in the Survey of India map as the Lung Tung glacier. There are two cols on this cirque, one to the southwest and the other to its north. Just beyond the ABC and to its northeast was a big glacier. This glacier, we discovered, goes along the northwestern aspect of the mountain that we wanted to climb. It is possible to go along the glacier and reach the head of the Rongdo valley. A route up the mountain was discovered from its western aspect along a southwest ridge which started just beyond our ABC. An attempt of the peak was made on 18 May 2013. The route up took six hours. After the first hour of climbing, we found the snow firm enough to use crampons. The route followed a north-easterly course. After another three hours of climbing, the slope

1. 'Climbing in the Rongdo Valley' by Joie Seagram in this volume.

became rocky and we had to take off the crampons. The last 50 m was like a tongue, a flat but steep slope. The peak had a two m cornice on the eastern side and we had to be careful not to step too close to the edge. On the way down, the right border of the tongue gave way and crashed down into the valley below. The ice on the track cracked and sank and we had to hurry across. This peak was named Sa'i Lhamo which is the Tibetan term for the Earth Goddess (*c.* 6030 m).

A team of three experienced mountaineers crossed the head of Tiburchan glacier and set up a summit camp to the north of a peak whose spot height on the map was 6160 m. This peak is located to the northeast of the head of Tiburchan glacier. A cirque located to the southeast of this peak and across the head of the glacier separates Tiburchan glacier from Koyak and Lung Tung glaciers. The cirque continues northwards across the head of the glacier, encircles the peak and then extends in a south-easterly direction separating the head of the Rongdo valley from Southeast Shukpa Kunchang glacier. The summit camp was a four hour walk from the ABC. Peak 6160 was attempted on 21 May 2013. A blizzard and poor visibility almost made the team sit out of the attempt. However they set out at 05.30 a.m. after the blizzard showed signs of abating. They took the northwest ridge up. The gradient was 45^0 but the snow firm and good for climbing. After three hours they reached 5990m. They had to then turn back as the weather deteriorated and fixing a rope for further climbing seemed unfeasible. This peak was named Lungkhor Kangri in view of the glacier encircling it.

Both summit teams returned to base camp and the entire group moved up the main valley. From the base camp the trail climbs steeply northwards for 20 minutes. On our way up this was still covered with a foot of snow. After a two hour walk, keeping to the right side of the Sakskyulas, we ascended steeply for 15 minutes. This brought us to a small meadow directly opposite the mouth of the valley explored by the Canadians. We established our new base camp- BC 2 here (*c.* 5170 m). The aim was to explore and climb in the upper reaches of the main valley. A recce team crossed Sakskyulas just below the base camp, skirted the terminal moraine of the Canadian valley and then followed the rivulet coming down from the left side of the main valley. Keeping to the right of this rivulet, they ascended a medial moraine and got a glimpse of the Southeast Shukpa Kunchang glacier at the head of the valley. There was another glacier coming from the east between

Lung Khor (Gazgazri) peak. (IAF 2013 Expedition)

Odgsal massif from Shan Lungpa glacier. Odgsal II (right) and Odgsal I (behind left) with Odgsal col in between. (IAF 2013 Expedition)

Ngapo Kangri and a pyramidal peak to the left of the Southeast Shukpa Kunchang glacier. This probably is the Sagtogpa glacier arising from the Kunzang group of peaks located to the northeast. A branch of the Sagtogpa glacier then swings south into the Rongdo valley. To the right of the Southeast Shukpa Kunchang glacier was a massif running in an east-west direction. This was separated from the Sa'i Lhamo massif by the Lungkhor glacier. The peak on the massif to the right of the Southeast Shukpa Kunchang glacier was the next target. To reach its base a summit team followed a rivulet coming from the right edge of the valley. After one and half hours they ascended steeply up the glacier at the head of the valley, keeping to the right edge of the valley all along. Then they crossed the Lungkhor glacier coming eastwards from R1 and established a summit camp at 5670 m below the peak marked as 6100 m on the map. The trek up to the summit camp took five hours. Hot sun and the fierce glare from the glaciers made it tedious. The team left for the summit at 03.30 a.m. on 25 May 13. The sky was overcast but there was no wind. The snow was crisp and firm. The team walked west parallel to the massif. After 45 minutes, they climbed up a ridge jutting southwards at right angles from the main massif. This brought them to the northern face of the mountain. From here it was a steep climb to the ridgeline of the main massif. The snow was firm and crampons made climbing easy. Once the ridgeline was reached, the snow thinned out and flaky rocks appeared. The gradient was less steep now and the climb continued in an easterly direction. The promise of a rose pink dawn was thwarted by clouds crowding in and a stinging blizzard. The team reached the summit at 07.02 a.m. (6183 m). This peak was named Khyung Kangri as it resembled the mythological Garuda[2] guarding the head of the Rongdo valley with its wings spread out. Khyung is the Tibetan word for Garuda. The massif is actually L – shaped and extends west – east along the glacier coming from R1 and then south – north along the right side of the Southeast Shukpa Kunchang glacier. The deteriorating weather made the team scurry back to summit camp and then further down to ABC the same day. The second summit team left ABC for the summit camp the same day. They planned to attempt a peak to the northwest of Khyung Kangri. They left summit camp at 03.15 a.m. but had to stop soon because of strong winds and heavy

2. A bird god, usually represented as a giant eagle, falcon, or part man and part bird. In Hindu mythology, Garuda carries God Vishnu through the air and attacks and devours serpents.

snow fall. After an hour they resumed their climb. They followed the same trail that the first team had taken the previous day. After crossing the ridge extending southwards from the Khyung Kangri massif they descended to a heavily crevassed area and had to rope up. They went in a northerly direction up a gully which went up the summit ridge located to the east. The snow was firm and they climbed using the three-point technique. After 200 m ascent, the route became rocky. They climbed a hump and could see the summit another 200 m above. A rope was fixed for 50 m as the gradient was 60^0 and the peak was summited at 06.17 a.m. This peak was named as Chu Skeyes Kangri (Chu Skeyes - lotus in Tibetan *c.* 6053 m). The team returned to summit camp at 08.00 a.m. and after a quick breakfast, decided to take advantage of the lull in the weather and attempt another peak nearby. The peak they selected was located to the southwest of the summit camp and was on a northeasterly flank of Sa'i Lhamo. The team took half an hour to reach the base of the peak and climbed it from an easterly spur. The team of six members reached the summit after two hours of climbing. This peak with its elongated easterly spur resembled the head of an elephant and was therefore named Langpoche (Langpoche - elephant head in Tibetan, *c.* 5969 m). The team then returned to the advanced base camp the same afternoon after a gruelling but extremely satisfying climb.

From the base of Khyung kangri looking east (R to L) Ngapo Kangri, Sagtogpa glacier, Pyramidal peak and Southeast Shupka Kunchang glacier. (IAF 2013 Expedition)

Head of Chhurn Nangma valley with Island peak in the middle and Odgsal massif to the right. (IAF 2013 Expedition)

The next plan was to go up the main valley, cross the Sagtogpa glacier, explore southeast Shukpa Kunchang glacier and climb the pyramidal peak between this and the Sagtogpa glaciers. However, the weather played foul. It snowed incessantly for three days. On the fourth day when there was a break in the weather we decided to descend to hot springs. The fresh snowfall had made climbing risky. We saw avalanches slithering down the peaks in the Canadian valley. Our attention now turned towards the Chudon nangma (Chhurn nangma in Survey of India maps) and the side valley further south.

After savouring the salubrious environs of the hot springs camp, a reconnaissance of the valley was done. It is a three hour walk on the left side of the valley till the valley closes in with a spur on the left side and rocky scree on the right. Two streams originate from the two sides and soon join up to form the river which we had followed up. The river was frozen at that time, covered by a thick icy layer about one m thick. We could hear the water roaring under it and the ice cracking in places, all intimations that summer was well underway. A group of four tried climbing up the spur on the left. There were two peaks above the spur which looked challenging. On going up the spur, they discovered a cwm there but the ice wall to the southeast and hanging glaciers made it too risky to climb from this side. Later we would discover that the best route up these peaks would be from the east across the glacier feeding the Chhurn nangma. Another team climbed up the rocky scree on the right side of the valley. From the top of the scree, a rock face was visible and a trail could be seen skirting the slope of the ridge to the right of the valley. A further scouting trip up this side revealed that after a two-and-half-hour walk, the valley suddenly widened and a huge glacial lake dammed by the rock face existed. This glacial lake is formed by the glacial melt of the Shan Lung glacier which lies like a crescent above the valley and to its south (Shan - snow leopard in Ladakhi), referring to the fact that we saw pug marks of the snow leopard all the way across the glacier. Straight ahead to the east was a big massif. A branch of the Shan Lung glacier swept northwards along the base of the massif and continued northwards east of the Canadian cirque. North of this massif was another glacier running in an east west direction and joining the Shan Lung glacier at the base of the massif. In the middle of this glacier was a solitary peak (Island peak) which looked interesting. But our first challenge was the big massif dominating the Shan Lung glacier. To climb this massif we established a new ABC 3 (*c.* 5305 m) just ahead of the rock face and on the banks of the glacial lake. From here, the route went eastwards to a small col (*c.* 5340 m) overlooking the Island peak. This was named Chugnis col (referring to the twelve-member Air Force team). From here, the route veered south-eastwards following the medial moraine of the Chugnis glacier to the base of the massif. Then the team skirted the base of the massif in an anticlockwise direction and reached a gully between the two arms of the massif. The massif has three peaks: the highest one on the north-eastern corner, the next one on the western limb, and a third plateau peak on a ridge extending southwards from the main peak. A narrow glacier descends in the gully between these peaks and goes on to feed the Shukpa Tokpo stream. A summit camp was established on a hillock in the gully (*c.* 5216 m). A team of four members and four Sherpas and Ladakhi nonoos left the summit camp at 04.00 a.m. The trail followed a ridge extending from the hillock in a north-easterly direction. The gradient was 45^0 and it took one hour to reach the head of the gully. We were now at the base of the col on the north between the two main peaks of the massif. We started climbing up the slope which had a gradient of 60^0 and 50 m below the col veered to the right i.e., in a south easterly direction. It took us two hours to reach the ridgeline from the base. From here it was a 30 minute climb to the main summit. The view from this peak was stupendous. There was a big glacial valley with a number of high peaks on the east (Muztagh glacier). To the west, the Shan Lung glacier and the Chhurn nangma were seen in their entirety. Across the Rongdo valley Karpo Kangri could be seen towering in the northwest. To the north the southern rim of the Canadian valley was visible and the Kunzang group of peaks to the northeast. This peak was named Odgsal I (Odgsal - clear light in Tibetan, *c.* 6234 m). The team descended to the summit camp and thence to the ABC in order to make place for the next summit team. This team followed the same route to the base of the Odgsal col and then rapidly climbed up the eastern face of the peak in the western limb of the massif on the next day. They were able to climb up in one and half hours. This peak was named Odgsal II (*c.* 6028 m). The third team was scheduled to climb the plateau peak (Odgsal III) but the plan was abandoned. Four days of bright sunshine had made the snow unstable and the route up western slope was fraught with avalanche risk. Instead another peak was selected close to the ABC. This peak located in the southern ridge of the Canadian valley was up the first gully ahead of the ABC. The team headed east from ABC up to the col and then veered left to ascend the gully in 45 minutes. They then ascended steeply up a glacier to reach a cirque in a little over two hours. This glacier was named Dabin glacier as it resembled a bowl (Dabin - means bowl in Ladakhi). From there they could see two peaks; one to the west and another to the east. The team decided to attempt the latter. The gradient was 70^0 and a

Approaching Sa'i Lhamo summit. (IAF 2013 Expedition)

Langpoche peak. (IAF 2013 Expedition)

rope had to be fixed for the last 25 m. The peak was climbed by the east-west ridge in five hours after leaving ABC. The Ladakhi climbers felt that the peak resembled a vulture head in profile and it was thus named Charok (Charok - vulture in Ladakhi, *c.* 6123 m).

While the third summit team was climbing Charok, the first team decided to go up the Rongdo valley up to the glaciers, a task they had left undone due to the bad weather. The team left base camp at 06.00 a.m. They first tried to ascend along the left side of the Sakskyulas but had to abandon the idea as the boulders made going difficult. They forded the stream and climbed on to the flat meadow which they had crossed while going to ABC 2. They then descended to the stream, forded it, skirted the terminal moraine of the glacier and the base of Ngapo Kangri and then turned left. Ascending sharply, they reached a snowfield which was the highest point (*c.* 5562 m) in the valley and reached at 11.00 a.m. From here, three glaciers were visible. The Southeast Shukpa Kunchang to the north, the Lungkhor glacier to the west and Sagtogpa glacier coming from the north of Ngapo Kangri in a westerly direction and then turning south into the Rongdo valley. The third valley on the left was also explored. Thara lungpa is a short valley dominated by a hanging glacier, Thara Kangri glacier. The glacier snout is 30 m high and one has to negotiate a snowfield similar to the one at Everest Base Camp to reach it. The climbers climbed along the right edge of the glacier snout and reached the glacier after three hours of climbing. There were two peaks above but the sheer rock faces made ascent from Rongdo side impossible. From Odgsal I we had already seen that these peaks could be climbed easily from the eastern side after crossing Shan Lung glacier.

The Air Force expedition spent five weeks in Rongdo valley. Summary of their climbs as follows:

Pk	Name suggested	Height	Date of Ascent	Team
1	Sa'i Lhamo	19785' 6030 m	18 May 13	Members:Gp Capt VK Sashindran, Sqn Ldr S Dubey, Sgt G S Pokhariya, Cpl JPS Raina, Cpl S Bohra, Cpl JR Poonia Sherpas/Ladakhi nonoos: Skalzang Rigzin, Tshering Bhutia, Karma Sherpa, Tashi Zangla
2	Lung Khor[3]	19086' 5990 m	19 May 13*	Members: Sgt C Y Basavraj, Sgt D K Vyas, Cpl D Chhetry Sherpas/ Ladakhi nonoos: Konchok Thinles, Tsewang Rigzin, Dawa Norbu, Stenzing Desal
3	Khyung Kangri	20286' 6183 m	25 May 13	Members: Gp Capt V K Sashindran, Sqn Ldr S Dubey, Cpl JPS Raina, Cpl S Bohra, Cpl JR Poonia Sherpas/ Ladakhi nonoos: S Rigzin, K Thinles, Tashi Gyalson, S Desal, Tsewang Gialston
4	Chu Skeyes Kangri	19860' 6053 m	26 May 13	Members: Sqn Ldr Avinash Prasad, Sgt CY Basavaraj, Sgt D K Vyas, Cpl D Chhetry, Cpl IA Dar Sherpas/ Ladakhi nonoos: S Rigzin, T Zangla, T Rigzin, K Sherpa, D Norbu, Cheder Bhutia, Dawa Gyalpo
5	Langpoche	19582' 5968 m	26 May 13	Members: Sqn Ldr Avinash Prasad, Flt Lt K S Sanghera, Sgt CY Basavaraj, Sgt D K Vyas, Cpl D Chhetry, Cpl IA Dar Sherpas/ Ladakhi nonoos: T Zangla, T Rigzin, K Sherpa, D Norbu, C Bhutia, D Gyalpo

3. Lung Khor (6160 m) was first climbed by the Canadian Expedition on 29 August 2013 and they named it as Gazgazri. Refer to 'Climbing in the Rongdo Valley' by Joie Seagram in this volume.

6	Odgsal I	20454' 6234 m		Members: Gp Capt VK Sashindran, Sgt GS Pokhariya, Cpl JPS Raina, Cpl JR Poonia Sherpas/Ladakhi nonoos: K Thinless, D Norbu, D Gyalpo, Tashi Phunchok
7	Odgsal II	19778' 6028 m		Members: Sqn Ldr S Dubey, Flt Lt KS Sanghera, Sgt CY Basavaraj, Cpl S Bohra Sherpas/Ladakhi nonoos: S Rigzin, K Thinles, D Gyalpo, S Desal
8	Charok[4]	20087' 6123 m		Members: Sqn ldr A Prasad, Sgt DK Vyas, Cpl D Chhetry, Cpl IA Dar Sherpas/ Ladakhi nonoos: S Rigzin, K Thinless, S Desal, T Bhutia

*Not summited. Actual height climbed is mentioned. The peak's height as per Survey of India map is 6160 m.

Summary

Detailed exploration of Rongdo valley of the Eastern Karakoram by a large Indian Air Force team in May – June 2013. They achieved first ascents of seven peaks and attempted one more.

4. Charok (6123 m) was attempted by the Canadian Expedition in August 2013 and they named it as Balden Lhamo (Rongdo III). Refer to 'Climbing in the Rongdo Valley' by Joie Seagram in this volume.

11

Kishtwar Kailash

Mick Fowler

In August 2013 the British Kishtwar Kailash Expedition was as ready as it could be. There were four of us, Paul Ramsden, Mike Morrison, Rob Smith and me. Our objective was the first ascent of the unclimbed highest peak in East Kishtwar and excitement levels were high. The troubles in Kashmir which have caused access problems for the last 20 years appeared to have subsided and Indian Home Office approval was expected any day now.

And then an Indian news report came to our attention:

12/8/2013 Srinagar News Report: In direct fallout of Friday's clashes in Kishtwar, all national highways leading to Jammu are now shut. Mobile and internet services were disconnected on Monday while schools and colleges remained shut.

A stressful month followed, the end result being that we didn't finally get all the paperwork in place until the day before we left. Once they have been overcome I like to speak positively of these challenges as they keep the crowds away and leave plum objectives unclimbed. But somehow I have trouble enthusing in this way at this time.

It appeared that the valley beneath Kishtwar Kailash had not been visited since Andy MacNae and Shan Singh Mann crossed the Muni la and travelled beneath the west face nearly 25 years ago. They had described the mountain as 'a very challenging objective.' And Paul and I already knew it to be an inspiring mountain having viewed it from Shiva the previous year and Cerro Kishtwar in 1993. Inspiring and challenging is a good combination we thought. And so it became our 2013 objective.

To avoid the troubles in Kashmir our approach was via the Rohtang pass and down the Chenab gorge. The last 50 km of road to the start of our walk-in, proved to be the most exciting that any of us had ever experienced. Paul took a three minute video clip of one section which had 835,000 viewings when posted on FaceBook.

The walk-in to base camp shared the first part of the once well-used Umasi la crossing from Paddar to Zanskar. The troubles in Kashmir

Kishtwar Kailash from north. (Fowler Ramsden Collection)

have hugely reduced the number of trekkers but conversely the 'Machail Yatra' (pilgrimage) in August, started only 40 years ago, now sees a remarkable 350,000 or so pilgrims covering the first two days to the small temple at Machail. The troubles in 2013 reduced that number by about half but even so that's a huge number of people to cater for, particularly on mountain mule tracks designed to link villages with no more than 200 inhabitants. But we were here in late September by which time the only evidence of their passing was a significant amount of rubbish and a curiously large number of well used portable toilets.

Beyond Machail our route entered the Darlang nala, a 70 mile long valley at an altitude of above 3500 m. It was immediately clear that we were on less travelled paths which had not been used by mules loaded with mountaineers' equipment for many years. Several sections had to be built up before the mules could pass and it was eight days after leaving the UK before we established a base camp a few hours short of Kishtwar Kailash at about 4000 m. What with the last minute permit stress and numerous uncertain moments with the mules it was a relief to arrive and be faced with just the challenge of climbing the mountain.

'Establishing a base camp' is actually rather a grand way of putting it. In practice each pair of climbers had one tent and Rinku, our liaison officer, Pritam our cook and Devraj the kitchen boy had another tent between them which doubled up as the cook tent.

The weather was indifferent but our time was restricted by full time jobs and every day has to be used to the full to stand a chance of success on a 6400 m peak in a 30 day trip from the UK. With such thoughts in mind Paul and I left promptly to explore and acclimatise around Kishtwar Kailash while Mike and Rob decided to explore a side valley.

Much as the cloud was down it felt excitingly adventurous to be heading up towards the peak that had been the focus of our attention for so many months. It would though have been better if we had been able to see it and thereby choose our preferred climbing line. Instead a full day was spent clambering over moraine covered glacier in the drizzle wondering exactly where we were.

On our third day out from base camp we reached a col at about 5700 m, still without getting a good view of the mountain that we had come to climb. The col had been spotted through clearings in the cloud and looked as if it would provide a suitable spot for relaxing, reading and acclimatising with maybe the possibility of ascending to a good viewpoint. In fact though it turned out to be a knife edge crest which, once flattened, gave a comfortable camping spot but little else. Fortunately though, after over 10 years of Himalayan climbing together, Paul and I are pretty well used to this acclimatising business and the fact that we spent the next two days pinned down in a small tent with just a small flattened area in front to walk about on was not a problem. We relaxed, read books and breathed in lots of thin air. At one point we even had a good view of the mountain that we had come to climb. We couldn't see the face we planned to go up but what we saw did at least convince us that our vague plan of descending the other side was a very bad one. If we were successful it was clear that we would have to descend by our line of ascent. After two nights in the tent on the little col we returned to base camp for a day of resting and eating and then we were ready to go. Meanwhile Mike and Rob had returned too and were fattening up in preparation for exploring the unexplored upper reaches of the glacier.

We don't bother with gadgets to deliver things like weather forecasts and so it was a surprise and relief to see clear skies as we waded the river by our base camp and left for the climb. Much as we hadn't been able to have a really good look at our proposed line clearings in the weather on the way down from our acclimatisation outing had at least given us a good idea of the best line to follow. Our plan was to avoid icefall danger at the base of the face by climbing a couloir leading to a possible entry point above the danger. Mind you to get there involved a lot of uncertainty in steep loose ground to gain a glacier, an icefall and a couloir of uncertain angle. The icefall looked easy from a distance but ended up being quite exciting with an overhanging section which required a heel hook to succeed. By the end of the second day though we had reached the edge of the face and were relieved to see that our hunch was right and we could easily access the face above the dangerous icefalls at its base.

It somehow feels more intimidating to step out onto a face and immediately be faced with exposure than it does to build up to the same position from the bottom. Here the climbing very soon became memorably atmospheric with smooth icefields and huge monolithic walls soaring above us and no obvious way through. But from my photograph, taken from the summit of Cerro Kishtwar in 1993, we knew that a fault line cleaved through the walls at one point and our hope was that this would provide the key to the lower part of the face. Classic European north face climbs tend to include sections of steep ice fields which succumb to calf muscle wrenching teetering on front points and initially the ground here was not dissimilar. As we gained height though, the ice became increasingly thin and the climbing increasingly precarious. By evening we were pleased and relieved to find a single spot to pitch our little tent on a small projecting prow just beneath the fault line. The weather remained glorious, the mountains beyond 'our' valley were increasingly coming into view and, as we snuggled down in the evening sun, it was beginning to feel that we were climbing in a very special place.

'Think I might have to come down'

Paul had already been up and had tried, removed his sack and tried again and now he was sounding uncharacteristically defeated.

'Round the corner might be better' he offered.

I hung from my ice screw belay, sucking in the cold morning air and peering around. There was another possibility to the right but it

Bivi 2, Rohini Shikhar behind. (Fowler Ramsden Collection)

didn't look any easier to me and to get there would require an abseil and a traverse that would take a couple of hours at least. And I really didn't want to do that. Losing upwards momentum can so easily lead to dithering and retreat. It looked to me that if we could just reach a point five m above Paul's high point the difficulty would ease for a few metres at least. Higher up we could see a clearly overhanging section of ice but that was not the immediate problem. The section causing difficulty was a loose and thinly iced bulge in the fault line splitting the huge walls. It had looked to be standard Scottish Grade V climbing before Paul had started up but now he was actually trying to climb it I could see that it was much steeper than it looked and clearly loose and unprotected.

'Shall I have a look?'

I had no great confidence that I could get up if Paul couldn't but somehow it seemed the right thing to say. There was a long silence.

Mick Fowler on icy slopes, day 5. (Paul Ramsden)

'I'll have one more go.'

As he inched higher and higher I felt a slight pang of guilt wondering if I had indirectly encouraged him to push on against his better judgement. This was no place to fall, there was no further protection and the climbing was obviously hard. I must have been almost as relieved as he was when he finally reached a belay and shouted for me to climb.

Paul Ramsden on headwall, day 5. (Mick Fowler)

As I approached Paul was looking relaxed and his usual positive demeanour was restored.

'Hard pitch that. Brilliant position.' he enthused as I gasped my way through inelegantly through.

The climbing for the rest of the day was not unlike some of the harder Scottish gullies with steep, sometimes thinly iced grooves interspersed with a couple of wild, overhanging sections. The millions of tons of snow that must have poured down here during the monsoon had compacted the snow such that climbing conditions were generally very good. Much pleasure was had and as evening approached another solitary projecting prow to one side even allowed us to pitch the tent. Life was good. Usually on this kind of climb we end up bivouacked with one person wrapped in the tent fabric and another in a bivouac sack, or sometimes we cut a bum ledge and sit side by side inside the tent fabric. On this route though we had managed to get the poles in the tent and pitch it after a fashion every night so far. And that makes such a difference.

Above us now the fault line we had been following reared up vertically for a long, long way to reach the summit crest just north of the south summit. Back in the UK our initial plan had been to climb the fault line in its entirety. However, before we left, close inspection of the various photographs we had managed to get hold of made it clear that

Abseiling on day 7. (Fowler Ramsden Collection)

Paul Ramsden leading ice groove section, day 4. (Mick Fowler)

the highest point was at the north end of what looked to be a potentially long and difficult knife edge ridge. And with that in mind our intended route ahead now lay leftwards following ice slopes under more huge blank walls to a shallow groove line leading up the headwall directly to the highest point.

We were rising above nearly all the surrounding peaks now and new horizons were opening. To the south the Prow of Shiva that Paul and I had climbed the previous year[1] was visible whilst the 7000 m peaks of Nun and Kun now reared their heads above the Cerro Kishtwar/Chomochior ridge on the far side of our base camp valley.

That evening we ended up on two separate 30 cm wide ledges hacked out of an ice patch on the headwall. Sitting there soaking up the evening sun high on an unclimbed peak and marvelling at the cloudless view was one of those 'it's great to be in the mountains' moments. As the last rays of sun left us I snuggled down in my sleeping bag using the tent fabric as a bivouac sack. Paul chose to sleep just in his sleeping bag. I seemed to be too excited to sleep but could hear Paul's slow and heavy breathing through the still night air. Away to the south a thunderstorm was brewing and gradually through the night it came our way. I lay there marvelling at the almost continuous lightening and wondering how close I should allow it to get before waking Paul and preparing ourselves for its arrival. But then it was 4.00 a.m. and our alarms were beeping so it was time to stir ourselves anyway.

On a bivouac like this it takes us about two hours from the alarm sounding to when we are ready to start climbing. Normally the first

At bivi 5. (Fowler Ramsden Collection)

1. See Note 'Prow of Shiva' in this volume.

Clearing site of bivi 4. (Fowler Ramsden Collection)

thing to do is get the stove going but on this particular morning some water must somehow have frozen inside it and the day started with Paul thawing it out down his trousers for 15 minutes or so. There was much relief when the gas started to flow freely and we were able to start melting snow for a cup of the best quality Yorkshire tea that Paul insists we have to have on these trips.

By the time we were ready to start climbing the thunderstorm had moved away but a front was clearly moving in. The way up the headwall was not immediately obvious but after first following a false line we gained a slanting fault line that gave some fine climbing leading directly to the summit. And what a wonderful summit it was. As I arrived Paul was sat on the knife edge crest with a leg on either side. The sun was still shining and with good visibility we could soak up the scenery and see that we were higher than any of the other peaks in this area. It felt a great privilege to be able to make the first ascent of a mountain like this. We loitered for 30 minutes taking photographs, generally revelling in the experience and contemplating that it's a wonderful feeling to dream of something for many months and then achieve that dream.

The weather stayed fair, the descent was pleasingly uneventful and seven days after setting out we were back at base camp. Mike and Rob had

Day 5, traversing towards headwall. (Fowler Ramsden Collection)

Mick Fowler on summit. (Paul Ramsden)

enjoyed their exploratory trip to the upper glacier, Pritam had prepared a Kishtwar Kailash success cake, relaxing felt particularly good and the enduring retrospective pleasure that goes with Himalayan success was beginning to flow. Long may the exciting peaks of Kishtwar be accessible for mountaineering.

The next day our perfectly timed weather window ended and snow came to base camp. The mules arrived as planned and it was time to go home.

Summary

Supported by Mike Morrison and Rob Smith, Mick Fowler and Paul Ramsden completed the first ascent of Kishtwar Kailash (6451 m) in alpine style from 4 to 10 October 2013 climbing 1500 m on the southwest face.

12

Climbs in The Hagshu Basin

Dan Kopperud

In late June of 2013 as our team took care of the final packing details, Uttarkashi in northern India suffered one of the worst floods in over a millennium. The towns and roads that linked the pilgrimage route to our mountain were washed away and wiped off the map. The mountain we had been training for and focusing on for 18 months was now off the table. Fortunately, Jake and Gabe were able to pour over maps and find a new objective for us. After hours of research and investigation they found a mountain in the newly opened Zanskar region of Kashmir. Our new objective was the 6515 m main peak of the Hagshu massif along with two other virgin peaks P. 6055 m (our readings on the climb showed it to be 6191 m) and P. 6035 m. In 2010 a French team attempted the highest point on the massif, but turned back due to unfavorable conditions. Team Everyman was finally reunited when we met in Delhi on 30 June. We immediately went to the library and checked out the maps to see what new route we'd have to acquaint ourselves with. Our initial reaction was shock and then excitement: the area we'd be going to had three unclimbed 6000 m peaks and there were many, many more in the immediate vicinity. We felt strong and ready to climb, so we refocused our attention on our new objectives.

Trek to base camp...

By the time we reached Kargil, the last city we'd see for six weeks, we were travelling in a small van crammed together. Our last stop was a collection of houses liberally referred to as a village named Akshow, just past Ringdom. On the morning of 07 July, we hired half a dozen massive, woolly yaks and donkeys to carry four weeks of food, climbing gear and video equipment.

We finally reached base camp on 08 July, which was set up on the last patch of grass high up the Hagshu valley. We acclimatised for a day by ferrying loads to advanced base camp (ABC) which was an arduous trek across rolling, constantly shifting and unstable moraines. On the morning of 10 July, Bryan, Jake and I went to ABC. We spent the next day fixing a line up a scree field which gave way to a waterfall that we'd be bypassing to get up to Camp 1.There was constant rockfall and we wanted to move quickly through this region. The next day, while Jean and Gabe started their move up to ABC, we left to find Camp 1. After traversing and climbing up around two more icefalls, we found a safer spot further up the glacier at approximately 5300 m, with fewer crevasses and better access to P. 6191 m and Hagshu I. This would be our permanent Camp 1. Here is an excerpt from Jake, regarding his thoughts on our journey over the last three days from ABC to Camp 1,

> The south-east ridge of Hagshu I rises in a clear knife edge thrust above a loaf of snow. It was at once stupendous, frightening and alluring. Two prominent rock buttresses guard the summit ridge and a thin cornice suggests, yet more challenge. The summit proper itself, tucked behind the rock fortress, is a massively corniced bump of snow and ice. By no means, does it look easy. It looks difficult, technical and long. Slowly, bit by bit, we are able to identify possible routes and ways thru each section. Break a big mountain down into smaller pieces that the mind can deal with. Think of the whole and you become…overwhelmed, intimidated, lost. The approach is finally over. Climb time is here. We must forge this culmination of dreams and sweat into dedication and (dare I hope?) success…

15 July, P. 6055 m, First Summit Attempt

After arriving at Camp 1 on 14 July, we decided to make a summit attempt that night. At 01.00 a.m. on 15 July Bryan, Jake and I hiked

P. 6055 m. (Jonn Jeanneret)

up around the rocky ridge located to the east of our camp. As we came around the ridge, we saw a couloir that looked like it would take us to P. 6055 m. We were surprised that it was only 30 minutes from camp, so when we hit the ridge, Bryan realised this was not the one we were aiming for. I was not feeling well, and was still feeling the effects of my efforts over the previous few days. So I decided to head back down. It was now about 03.00 a.m. and they saw a couloir just around the cirque and headed over to start climbing its lower flanks. This portion of the couloir was simple 50-60 degree snow and ice, so they soloed up. About halfway up the couloir they ran into some overhanging seracs. Bryan led the ice pitch which ended on a beautiful ledge overlooking the entire cirque. Jake ran out the full rope length on the next snow pitch, but ran into vertical alpine ice just above his snow anchor. They climbed the next 200-300 m of ice 60 m at a time.

At the top of the ice they found a great little cutout in the snow to rest and get some food and water. They climbed a 15 m ice pitch at the north end of the cutout to gain the ridge, but at this point a massive whiteout had consumed the entire upper portion of the mountain. The wind was howling so much that they weren't sure if it was snowing or just blowing up from below them. They could see about 100 m ahead and noticed a large rock up on the ridge, only about what appeared to be 100-150 m in height, from the summit. They decided to head for that rock and take cover for a few hours. It was only 11 a.m., so they assumed they had some time before they needed to head back down. Bryan grabbed the rope and headed straight toward the rock ridge, when he got there he was amazed to find another cutout similar to a moat in the front of the first rock ridge. What was different about this cutout was that it contained an alpine lagoon. Bryan cracked the ice and filled his bottles with some cold but well-needed water. Up on the snow the wind was ferocious, but down in our moat it seemed like a calm fall day with the temperature cold but manageable. The boys had now been out for 12 hours and were in position to summit, if the weather would simply clear enough for them to see.

At 04.00 p.m., Bryan set out again for the summit, but after only 100 m, he couldn't see where to head next. Knowing that this area was possibly heavily crevassed with a few hanging seracs overhead, Bryan realised it was time to head back to the ravine and wait. By 05.30 p.m. the storm hadn't let up and with dark coming they realised they couldn't wait any longer and decided to head down. In the meantime, Jean and Gabe had joined me at Camp 1 and we waited for Bryan and Jake to return. When I came down I realised that I had the summit team radio, so communication from the mountain was impossible. It was now after midnight and the guys had now been out for more 24 hours straight, so we started to worry. Meanwhile….

> We chose to risk a night descent of the 70^0 slopes. We began to reverse our morning's free solo – down climbing into the white gloom guided only by the few metres of footprints we could still see. A less tired soul might have been scared, but we accepted the odds and worked our way mechanically down. Darkness came on and we switched on our headlights. Two pale arcs of dying dim light, less than that of a small birthday candle was all we had. Barely able to see our own feet we clumped ever down. Finally we emerged at the bottom of the slopes, exhausted. We roped up and more by instinct and experience than by corporeal senses we avoided the gaping crevasses in the dark night and trudged the kilometre or so back into camp. It was 03.00 a.m.; we'd been out 26 hours.
>
> - excerpt from Jake's diary

Jake and Bryan had now returned out of food and water and were extremely tired. 'The route will go,' were their last words, as they crawled into the tent, desperate for some shuteye.

Our main objective was still Hagshu I, directly west of Camp 1, so the next day, Jean and I climbed up the glacier to the base of Hagshu to scout for a possible location for Camp 2. We found a suitably safe place in the flat expanse between P. 6035 m and Hagshu I in the middle of a bowl surrounded by enormous, unclimbed mountains and eagerly came back down to Camp 1 to report our find.

P. 6035 m, Summit Attempt

We all moved up to Camp 2, on 18 July, and made a plan to attempt P. 6035 m just after midnight. On 19 July, at 01.00 a.m. we left in two teams. Bryan and Jean were on one rope and Jake, Gabe and I were on the other. Jake describes the morning quite well.

Bryan's cheery 'Who wants to go climb an unclimbed peak?' rang out just past midnight, echoing across the hoar frosted night air and bouncing off the cirque of ice and rock that bound us here in our orange and green beetle-esque squats. Spirits rose with our stinking bodies and as fast as possible we donned gear, roped up and set off into the still inky-blackness. The almost nonexistent airy bridge over the bergschrund was reached and breached. Two pitches of ice later, a growing steepness of ice and eagerness to lead from Dan, saw me happily surrender the lead. Some anchors had been more hope than hold, and excavations of 30 cm to find solid ice frustratingly common. Still we were moving well and the climbing was thoroughly enjoyable.

The grey wave of pre-dawn light revealed a jagged horizon to our left, then it turned egg yolk yellow then pink and gold as the sun breached the curve of turning Earth. In minutes the kiss of cold beams touched us and the steep slopes that arched away above us. Soon the U.V.s would begin the treacherous softening of snow and anchors. Then illness struck. Vomit, shit and phlegm erupted in violent fits from Gabe. Jean and Bryan were now too far ahead to offer help and besides, we had no desire to halt their progress. A few minutes was all

P. 6035 m East Face. (Jonn Jeanneret)

P. 6035 m North Face. (Jonn Jeanneret)

> it took to make the heart wrenching decision. We needed to go down. Dan did an excellent job setting V-threads in the hard to find ice (when there was ice!) and together our team of three negotiated each pitch and peril back down and over the 'schrund to camp 2.

In the meantime, Jean and Bryan were still high on P. 6035 m. The next two pitches to the summit ridge were exciting and exposed. Jean was starting to struggle and seemed to be losing the battle with his cough. On Bryan's first step, above the ridge plateau, he punched through to his waist and could see air below his feet. It was time to find a new route.

They wandered around the backside of the summit ridge and noticed the summit we were heading towards was a false summit. From our Camp 2, the summit was hidden behind what we thought was the peak's high point. They walked around to the southeast face and located the true peak. While Bryan brought Jean up he noticed an opening in the cliffs below, with a snow ramp down the middle that led to the glacier. It would be a walk off option for a descent, but would put them on the exact opposite side of the mountain from our camp.

At 01.00 p.m. they reached a small rock ledge just below the summit. The final pitch was a mix of snow, ice and rock. Bryan put Jean on belay, confirmed with him how to properly place an ice picket and screw then he set off. Jean then proceeded to hack every bit of snow and ice off of the face and rock climb the final pitch. He made the summit, screamed aloud and filmed a few short videos. Bryan followed using the freshly exposed rock and had no problems finding Jean's route to the summit.

At this point, a storm had started rolling in and it prevented them from seeing a descent route down the west ridge and since they didn't have the v-thread tool they couldn't rappel back down the face they came up. Instead, they took a snow ramp option that Bryan had seen on the south face during the ascent. It meant they could possibly walk off, but it put them on the opposite side of the mountain. They would need to circumnavigate the east side of the peak. Since there was no other option, they headed down and in a complete white-out finally made it to the glacier.

It was now about 08.00 p.m. and the adrenaline of the summit and down climbing was wearing off. Every step was hard work. Not far from camp, Bryan started to feel the psychological effects of the all-day climb. They saw no head lamps or sign of tents and their calls were answered by only the howls of the blowing wind and the fury of the storm. As they laid down one final time in the snow they saw a familiar crevasse crossing and knew that they were close! Bryan saw the dark shapes of the tents so he called out but got no response. Not until they were within 10 m of camp, did they finally hear voices and see lights come on. They made it! The guys named the route, 'Under Moonlight'.

Camp 3 and Hagshu I attempt

After a rest day and more bad weather, we started making plans to climb Hagshu I. It sits on the opposite side of the glacier from P. 6035 m facing Under Moonlight, the route Bryan and Jean climbed. We could see what looked like a low-angle shoulder on the ridge of Hagshu and we planned to make Camp 3 there. Bryan, Jake and I left at 05.00 a.m. up a couloir that led to an ice face below the shoulder we planned to establish camp on. We spent the morning fixing lines and finally traversed around the mountain to the ice wall. Bryan led through the ice while an exhausted Jake and I followed. Here's the same day in Jake's ever so eloquent journal entry…

Hagshu I, face attempted

The push up to Camp 3 was long, hard and brutal. Burdened like human yaks (and with our scruffy beards and unwashed bodies, the analogy is closer than you might imagine!) with 25 kg or more Bryan, Dan and I zigzagged our way up the lower slopes. Ever above us rose Hagshu 1 and the mighty rock buttress, which seemed to guard the way to her summit. It hovered there over us, a physical and mental forbidding menace. We crested the first col and fixed some 220 m of twisty, recalcitrant, yellow nylon rope. When this ran out we began pitching on our last remaining 60 m rope. Vain promise, after illusion, after false promise of the ridge crest forced us ever up in a Sisyphean labour. As darkness threatened its arrival, Bryan's joyous curses wafted down to us at our pitiful anchor. Camp 3 had been found. It was a narrow rock promontory that jutted out from the surrounding slopes like a shrugged boney shoulder or minor dorsal fin belonging to

> some peculiar creature of the deep. The little two-man tent sat awkwardly on a rough leveling of rock. On all sides the mountain dropped away in vertiginous plummeting sweeps of either rock or ice. Only to the back, in a narrow soaring hunch did it rise above us and towards that dark buttress. With snow and wind snapping at the three of us, we crammed into the tiny tent. Three puffed up sardines soaked in sweat and squeezed into a supposed two-man orange nylon can.

The next morning Jean arrived and we set up the tent he brought, sorted our gear and made a plan for summiting Hagshu I. The ridge ahead of up was low-angle snow that led to nearly vertical ice which stopped at a gendarme, beyond which more ice and a rocky, exposed summit ridge. That night we were blanketed in some new snow which did not let up the entire next day, so we were all forced to wait another day, huddled in our tents, perched high up on Hagshu's shoulder.

At 04.00 a.m. on 25 July, we started climbing unroped up the gradually steepening ridge. Jake didn't have a good feeling about the conditions and decided to turn back. Bryan, Jean and I continued up the ice. The fresh snow covered the cracks and crevasses that had been becoming more exposed as the summer wore on. At several points we dug through snow to find solid ice for placing screws only to find rotten ice or massive empty expanses. We placed bad gear or no gear at all in these situations and continued up, making sure to make solid tool placements where we could.

As I was topping out the last pitch, I noticed what looked like a piece of blue cord. After I built a sketchy anchor out of four equalised snow pickets and a bird beak I looked up to inspect the cord. About five metres above me on the gendarme was a piton with blue cord sling through it. It appeared as if we'd encountered fixed gear from the French team who was here in 2010. We decided to skirt the gendarme and climb it from the right side. Jean started dry tooling and aiding up the rock before he got to the ice. We needed to decide our level of commitment and acceptable risk very quickly as the clouds enveloped us again. The ice had formed in a curtain where it was attached only at its top. We'd need to traverse it, with no protection, in order to get across the chasm to the next gendarme. None of us were willing to lead through it, especially with an impending whiteout looming. We

Hagshu I (6515 m) from east. (Jonn Jeanneret)

had to face the reality at this point that if we didn't continue, our shot at Hagshu I was finished. We all accepted this realisation as we made our high anchor to rappel down from. Several hours later we were back down to Camp 3. The next morning we broke camp and made our way back down to Camp 1 on the same day.

P. 6055 m, First Ascent

Bryan, Jean and I wanted to make one last attempt at P. 6055 m. So, on 28 July at 01.15 a.m. we left under perfect conditions. The stars were out and they lit our path as we climbed the initial snow slopes. We simul-soloed until we got to a frozen waterfall that Bryan had climbed during the previous attempt 12 days earlier with Jake, but now it was melting out and had become radically different. The pitch was a wet, flowing waterfall covered by a foot of solid ice. At the top of the pitch we decided that it would be fastest if we simul-climbed the 300 m ice wall to a recess in the snow. We continued unroped up the snow slope and as we climbed through the couloir. We unroped above the ice and soloed the final 250 m to the final pitch of technical snow and ice.

We weren't sure what lay above or what the summit would even consist of. As Bryan climbed, he paused to dig snow away to find solid ice below to place protection. As he got higher, the ice became rotten and wouldn't hold a screw so he placed pickets whose inevitable function was little more than lightening the weight he had to climb with and some psychological security. He pushed on through the steep snow, kicking down sloughs that landed on Jean and me with every step.

'It's all soft snow!' Bryan called down to Jean and me, 'I can't get any gear in!' He was now two metres below the corniced ridge and six metres above his last, useless picket. He didn't have anywhere to place gear and at any point could come hurtling down on top of Jean and I. The cornice was below what appeared to be either a summit plateau or a drop into space and Bryan was out of pickets. He had two screws left but there was nothing that would take a screw so he set his feet as solidly as he could and gingerly pulled on his axe to not disrupt the snow they floated in. He stood, on his precariously balanced feet and poked his head over the ridge. We heard his signature 'Whoop!' and we knew he'd made it. A wave of relief and excitement swept over me. Jean and I followed and met Bryan on the slightly angled summit plateau where he'd sunk his last two screws. We looked around and noticed the hourglass shape of the summit and the two independent peaks of equal height. We checked the altitude and it was 6191 m. Apparently the survey done several decades before was a little low.

We took in the view around us and saw for miles in every direction. Finally, we had a clear day for a summit attempt and it was gorgeous. Below us was the Hagshu glacier and directly across from us was Hagshu I, with Camp 3 as high as we were now which served as a reminder of how huge Hagshu is. The sheer vastness of the area was awe-inspiring. We took our obligatory summit shots and Jean recorded a video proposing to his girlfriend, Brenna, from the summit. It was a bittersweet moment when I had the realisation that this was the culmination of 18 months of training; that in a few minutes we'd be heading down.

This region was opened officially for climbing again in 2010 and there are so many new objectives that climbers haven't begun to explore yet. There is a bright future for climbing incredible peaks and faces in Kashmir. It's a little overwhelming but mostly inspiring to think about the all possibilities and potential that this area holds for generations to come.

Oh, and Brenna said 'Yes!' to Jean's proposal.

View from upper slopes of Hagshu I. (Jonn Jeanneret)

Team Members: Bryan Hylenski - Leader, Jake Preston - Medical Officer, Jonn Jeanneret - Photographer, Daniel Kopperud and Gabriel Thomas.

Note by Lindsay Griffin, Senior Editor, American Alpine Journal:

Hagshu has confusing history of ascents. The peak has been reported to be around 6330 m, but the Survey of India marks it as 6515 m making it possibly the highest in the east Kishtwar region. The same map shows a distinct north top (around 6330 m), which sits above the much-attempted north face.

In 1988 a Polish expedition approached the mountain from Zanskar to the north. They ascended the glacier below the east-northeast face and eventually reached the plateau at c. 5700 m below the southeast ridge. In 1989 they returned, but this time from Kishtwar to the south. Due to consistent bad weather, most of the team eventually decided to abandon the attempt, but Pawel Jozefowicz and Dariusz Zaluski opted to stay. The day after the others left, the weather cleared. The two Poles climbed through a crevassed area and then a 500 m couloir (up to 60°) to reach the plateau below the southeast ridge. From here they climbed the southeast ridge in two days, reaching the summit on 9 September;

there was a rock step of UIAA VI- and a 15 m ice step of 80°. They descended the same way and walked out to the north, arriving at the base camp of a British party. As the Poles did not have a permit, this ascent remained undercover.

The first official ascent was made just one week later, on 16 September, by Phil Booth, Max Halliday, and Ken Hopper (U.K.), who approached from Zanskar. They climbed a steep glacier system to reach the serac-torn east-northeast face, where they climbed a line up the centre with two bivouacs. They took a day's rest at each bivouac site. The hardest climbing was above the second bivouac—a couple of pitches of Scottish 4. They followed the summit ridge southeast to the highest point (1200 m, TD, with much 55° névé).

Hagshu saw multiple visits by John Barry and various partners attempting the impressive north face; probably the best effort was in 1994, with Seb Mankalow, reaching c. 6000 m. After nearly 20 years, the peak was attempted again in 2010 by a large French expedition, which made little progress on the mountain.

Summary

In July 2013, a team of expatriates from Korea made two first ascents on P. 6035 m and P. 6055 m (their readings came to 6191 m) and also attempted Hagshu I (6515 m) up to 6440 m from where they had to return due to poor ice and weather conditions in the Hagshu basin of the Zanskar Himalaya.

13

Portals to the Past

Trekking in Upper Mustang

Lakshmi Ranganathan

On my Annapurna Circuit trek way back in 2003 as I descended to Kagbeni located on the banks of the Kali Gandaki I saw a board at the edge of the town that read – 'STOP- You are now entering the restricted area of Upper Mustang'. Looking beyond I saw waves of seemingly barren brown and ochre scree slopes. What caught my eye though was that some of them had dozens of high close set caves and I wondered if they were natural or man-made and whether they were inhabited. There did not appear to be even a pith of a path leading to them. Upper Mustang was again brought into my consciousness when I read Michel Peissel`s fascinating book[1], where the picture of the author`s landlady at Lo Manthang gave me the creeps.

Early history

Pronounced 'Moos-thang', meaning fertile plain, Upper Mustang constitutes about two third of the Mustang district and was opened to trekkers as late as 1991. Lower Mustang was well trodden though housing the famous Hindu shrine Muktinath and part of the Annapurna Circuit trail. With records of its existence as early as the 8th century, Upper Mustang was originally a part of the far western Ngari region of Tibet. Untouched by the political travails of mainland Tibet by the Chinese invasion and the Cultural Revolution it remained one of the last bastions of true Tibetan culture and lifestyle. Ame Pal (1380-1450) is considered the founder king of a united Mustang kingdom. Around 1795, after the Gorkha kings came to power and conquered the adjoining kingdom of Jumla, by a strange twist of historical fate or fortune depending on how one looks at it, the kingdom of Upper Mustang was also peacefully and loosely annexed by Nepal. The Mustang Kings however had considerable autonomous power, ruling the province from its capital, Lo Manthang (Plain of Aspiration). The present much loved and respected 'King', Jigme Dorje Trandul, also known by his Nepalese name Jigme Parbal Bista still resides here and is said to be the 25th in a royal line that began with Ame Pal.

1. *Mustang - A Lost Tibetan Kingdom* by Michel Peissel, Book Faith India,1967

The Government of Nepal however ended the monarchy as late as October 2008 the year it became a republic.

Due to its strategic location Upper Mustang was an important outpost on the salt and wool trade caravans between Tibet, Nepal and India. After the Lhasa uprising and Dalai Lama`s escape to India in 1959, this border was sealed and Upper Mustang turned into the hideout of the Khampa soldiers who were secretly trained in modern warfare by the CIA directed against the Chinese. However after Henry Kissinger`s secret visit to China 1971, paving the way for Sino-American treaties and China and the Dalai lama putting pressure on Nepal to disband the secret army network, Upper Mustang gradually opened up. Nevertheless it had its fair share of visitors in the guise of scholars, historians, explorers, and surveyors since 1868[2] who entered its portals either ceremoniously or surreptitiously, notable among them being the Indian pundit-explorers, W.J. Kirkpatrick, Ekai Kawaguchi, Sven Hedin, Toni Hagen, David Snellgrove, Giuseppe Tucci, Herbert Tichy, Peter Mathieson, Thomas Laird and Michel Peissel. In 1999, the construction of a motorable 'Trans Himalayan Highway' on this ancient trading route right from Kore la on the Nepal- Tibet border to Jomsom began, thereby invoking the quip –'Now you can ride from Katmandu to Mustang in a Mustang'! The same year also saw the escape of the Karmapa from Tibet to India by this very same route. Tucci appropriately described this route as 'used over the centuries by pilgrims and apostles, robbers and invaders'.[3]

In Sep 2012 two friends and I decided to trek to Upper Mustang after waiting nearly for a decade with the futile hope that the exorbitant restricted area fee would be either reduced or waived! Another fear was that the new road would render the trekking trails defunct. It was a sad sight to see small road signs marked with arrows on the now largely motor Annapurna Circuit trail that read 'Original Annapurna Circuit route' and before we saw similar signs in Upper Mustang, we decided to hit the tracks. The wait had its reward though as it was now possible to do a light weight basic tea house trek instead of a bulky camping trip. The news that one could also visit some recently 'discovered' ancient

2. 'The Survey of India and the Pundits' by Michael Ward, HJ Vol. 55, p. 99
3. Trekking in the Nepal Himalaya, *Lonely Planet,* Edited by: Mayhew, Bradley and Bindloss, Joe, 9th Edition, Aug 2009

caves was an added lure. We used the services of a trekking agency at Kathmandu to arrange for all the paperwork and other logistics. We had with us Mek Bahadur (guide) and Bimal (porter).There was however one unexpected surprise.

Day 1: Jomsom-Kagbeni-Tangbe-Chhusang-Chele

Our trek started at Jomsom into which we flew from Pokhara. The next stage, Kagbeni, still had its charming old world ambience with its protector deities. Although the trek route largely followed the motor road, the numbers of vehicles were few and far between, thus making it a pleasant walk. We ascended on a high trail along the eastern bank of the Kali Gandaki slowly sinking in the spectacular scenery that unfolded at each bend even as the typical noon wind blasted grains of sand on to our faces. The mysterious caves, the holes in the hills like numerous eyes were everywhere as if watching us. Complementing the brown slopes were the dazzling white Nilgiri North, Central and South peaks towering above Kagbeni in the southern horizon. Ruined forts and dwellings on the crags and red, white and grey coloured chortens typical of the Sakya Buddhist sect that largely prevails here dotted the landscape.

Painting in Tashi Kumbum. (K.Mahesh)

Protector deity in Kagbeni. (K.Mahesh)

Soon a green patchwork heralded the village of Tangbe which has an organic apple farm on its outskirts. Just after Tangbe, the trail led through a stunning section of the Kali Gandaki canyon whose fluted red and orange 'organ pipe' walls were pitted with numerous caves. The next village Chhusang lay below these cliffs at the confluence of the Kali Gandaki and the Narsing khola. As we walked past its narrow alleys and mud plastered houses we truly felt transported back a few centuries. Crossing the Narsing khola over a log bridge the trail continued along the river bank. The canyon soon narrowed just below the village of Chele where one channel of the Kali Gandaki flowed through a narrow tunnel formed in a talus cone that has broken off from the cliff. A pedestrian steel bridge also spanned the river near this tunnel and after crossing it, the trail climbed high to Chele where we halted for the night in a lodge. Although the jeep track ends currently at Chhuksang, it is possible to drive across the river to Chele when the water level is low as in winter.

Day 2: Chele - Samar - Chungsi Cave - Shyangmochen - Geling

Chele, with its wood stacked terraces and goat-head skulls and spirit traps adorned houses and chortens and mani walls at different levels truly has a medieval mien about it. We walked out of it to a large steel suspension bridge spanning the deep canyon walls of the Ghyakar khola. We did not cross this bridge but ascended a steep narrow scratch of a trail along the very edge of this canyon with great views of the picturesque village of Ghyakar across. Past two passes Taklam and Dajori la, the trail descended past mud walled fields to the historic village of Samar, the hotbed of the CIA trained Khampas in its heydays. From Samar the trail forked and we took the lower rough trail that switch backed through two side valleys of Kali Gandaki, the Samarkyung and Jhuwa kholas to reach a high pass, the Chungsi la. The descent from here to the Chungsi caves was spectacular as the sandy trail contoured through narrow pleated canyon walls with amazing rock formations pitted with giant caves. The sacred Chugsi cave was well hidden deep in these walls and its main attraction was a giant 'self-risen' stalactite in the shape of a chorten (according to the Buddhists) that is said to be still growing.

After Shyangmochen we were back on the high road. Just above the village from the road pass, Shyangmochen la, the highest point of the day we got a panoramic view of the Annapurna range and I was surprised to see a vaguely familiar giant of a mountain, besides the Nilgiri and Tilicho peaks. It was none other than the north face of Annapurna I, with its prominent ice cliff that Maurice Herzog had called 'The Sickle'. 'The Grand Barrier', that massive wall of cliffs that confused the French climbers in 1950 and blocked the entry to the base of Annapurna I was also seen in its full fluted glory. This visual treat was so unexpected and I knew that it could only get better as we trekked further north.

Past a giant chorten, the sprawling green oasis of Geling came into view with a red gompa high above the village. Taking the right fork (the left fork was a direct route to the next major village, Ghemi) we descended to the village.

Day 3: Geling - Ghemi - Dhakmar - Ghar Gompa

We visited the Geling gompa on our way out of the village. Women were not allowed inside as it had a blackened chopped palm, presumably belonging to a thief. However the views from here of the Annapurna range were great. We gradually ascended above Geling alternating between the road and short cuts to top two passes Nyi la and Ghemi la where the views especially of the Grand Barrier got grander. Soon verdant fields heralded the village of Ghemi in the valley of the Ghemi khola and towering above it were stunning deep red streaked hills.

Ghemi, with its historic mani wall was the third largest village of Upper Mustang after Lo Manthang and Tsarang. Two routes lead from here to Lo, the motorable one was via Tsarang and the other roundabout but more vivid trail was via Drakmar, which we took. We headed north past the longest mani wall in Nepal, about 200 m in length. From here to the next village Drakmar (meaning red cliff) was an incredible trail of flame red convoluted conglomerates with its caves and crannies that towered over green fields and willow trees. Crossing the bridged Drakmar khola we ascended steeply above the village to the Mui la from where more views of the snow peaks and 'ripped' weathered crags could be had. On our descent and traverse to the Ghar gompa we could see Tsarang on the eastern horizon snuggled below slopes that strangely had a giant right footprint pattern on it.

Ghar gompa, situated high above the Lo Gekar (Saukre) village was an 8th century Nyingma monastery. Legend has it that this monastery had to be constructed first by Guru Padmasambhava to subdue the demons that were preventing the construction of the Samye monastery; the oldest one in Tibet.

Dhakmar village. (R.Lakshmi)

Gangapurna (extreme left), Annapurna 1 (centre),Tilicho and Nilgiri peaks seen enroute Kore la (K. Mahesh)

Day 4: Ghar Gompa-Lo Manthang / Around Lo Manthang

Descending from Ghar gompa we crossed Ghyung khola and ascended to the Choga la. A gradual descent past the ruined forts of Samduling brought us to a vast plateau. Below infinite 'silky' beige and purple ranges, past irrigation channels and fields appeared the fortified walled capital 'city' of Lo Manthang. About 150 houses are contained within its gated, L shaped walls, 750 m long and nine m high with 14 towers. Very Tibetan in culture, its people are known as Lobas and the the walled city is very intriguing. We visited the intricately frescoed Jampa, Thubchen and Chode gompas and the King's palace. Ascending north to the village of Chhon up from Lo Manthang, we saw two small sandy hills looking like volcanoes both of which had ruined castles on their peaks. The higher one was the historic Ketcher Dzong[4] the castle of Ame Pal, with its oval shaped ramparts. The lower one was that of his queen, which is now used as a burial site. The views from this castle were incredible. In the vast plain that stretched below us with its patch worked fields, encircled by a chain of peaks was, in the western side, the King`s summer palace at Tingkhar village, Phuwa and Kimaling villages with ruined castles and the Namgyal village with its gompa and southwards was Lo Manthang dwarfed by the Annapurna range.

Day 5: Lo Manthang - Kore La – Nyiphu - Jhong Caves - Garphu - Lo Manthang

Setting off early in a vehicle, we headed north of Lo Manthang on the very bad non metalled road leading to what is called the Chhosar

4. 'The Mustang Beat: Two unheeded historic sites' by Phillip Sturgeon, *HJ* Vol. 56, p. 56

North face of Annapurna I. (K.Mahesh)

valley along the Nhichung khola, a tributary of the Kali Gandaki. On the eastern horizon across the river were the villages of Nyiphu and Garphu abounding in cave dwellings. All along the road were chortens, ruined forts and other crumbling structures of yesteryear civilisations. Fording the Nhichung khola we started ascending steeply. As we ascended higher we could see the entire Annapurna range rise in its fully panoramic glory. Crater like summit ridges rose to the left of Annapurna I. The summits of Annapurna II, III and IV, apart from Gangapurna and Khumjungar and other peaks were seen. Just before the Kore la welcome arch (4400 m) was one of the stunning viewpoints from where all the four Annapurnas could be seen in one frame. Soon the steep ascent gave way to a flat traverse across a broad desolate pasture land. In another 10 minutes we were at the Kore la border (4660 m) where we saw a long fence with a locked gate at the roadhead stretching for kilometres on the vast plain. A border pillar, # 24, 1962 with Nepal and China written on either side stood on the Nepal side. This plain is called Ghoktang, the terminus of the 102 km Jomsom-Ghoktang road. Seasonal trading takes place here regularly between Mustang and China accounting for the deluge of Chinese goods in the whole of Mustang[5,6]. We were told by the locals that although Nepali vehicles or people were not allowed to cross that fence, the Chinese regularly opened the gate and drove all the way to Lo Manthang, hung around for a while smoking their cigarettes and then went back.

On our return, we forded the Nhichung khola twice to reach the village of Nyiphu with its cave gompa. Close to it were the more intriguing Jhong caves. We were now actually entering the 'caves in the hills'! The Jhong caves system was about 2500 years old, with five stories and 44 inter connected rooms. These caves like many others around here were once part of an ancient civilization. One postulate suggests that these caves were initially used as funerary chambers, which is supported by the discovery of human skulls and bones in some of them. Later as wars and invasions became commonplace, they were used as dwellings as safety became a prime concern. Later in more peaceful times, coupled with the lack of water forced people to move into valleys while the caves themselves fell into ruins or gradually transformed into places of worship, storage chambers, graves and military hideouts.

First view of Lo Manthang. (K. Mahesh)

---

5. 'The Road to Lo' by Abisek Basnyat, *Nepali Times,* Issue # 159, August 2003.
6. 'Monsoon in the rain shadow' by Kunda Dixit, *Nepali Times,* Issue # 621, September 2012.

Days 6, 7: Lo Manthang – Dhi – Yara / Yara – Luri Gompa – Tashikumbum - Yara

We heard about a deep canyon trail to Dhi that led southeast of Lo Manthang along a hill called Udi Danda that sounded too good to miss. As we skirted the range we could see the road deep below and the Annapurna range in the south. We descended steeply into the dry canyon where at some places the towering walls were less than six m apart. Just as we came out of it, the terrain opened out suddenly to reveal the verdant valley of the Mustang khola, as the Kali Gandaki is called in this region with the villages of Dhi and Surkhang on its true north and south banks respectively. The petite village of Dhi looked crushed under talus columns that towered above it. Crossing the Mustang khola the trail led along the banks of a tributary, Puyon khola for a short while before we climbed steeply to the village of Yara. From Yara a high ridge trail led to the village of Gara and further on to the Luri cave gompa. We had great views of the imposing north face of Dhaulagiri and Tukuche peaks from Gara. The 14th century cave gompa belonging to the Drukpa-Kagyu sect was effectively a retreat centre for the monks of the actual Luri gompa situated a little below the trail. From here a trail led further east to the Hindu pilgrimage site of Damodar Kunda, a

'Organ pipes' scree slopes near Yara village. (K.Mahesh)

group of small lakes. While mythology claims it to be the source of the Kali Gandaki, a major tributary of the Ganga, some reports (Wikipedia) indicate its source as the Nhubine Himal glacier, west of Kore la.

From Luri gompa, a Buddhist nun reluctantly escorted us to the Tashi Kumbum, another cave gompa that was not open to the public. Following a different route back to Yara, along the Puyon khola river bed was an indiscreet looking crag with many caves about 50 m above the river. We had to scramble to reach the locked door at the entrance of the cave. Just as in the Luri, here too was a chamber with a vandalised broken chorten, extremely beautiful wall and ceiling paintings, tsatsa moulds and scattered scriptures in Tibetan script.

Tashi Kumbum is one of the many cave gompas that are being discovered either accidentally or assiduously housing valuable historical artefacts. Notable among these are those found near Samdzong and Konchogling in the Chhoser valley. Many discoveries are now being hushed for fear of robbery. What was intriguing about these paintings was that most of them did not have a Tibetan imprint but were more aligned with the Indian style of paintings seen in the Ajanta and Ellora caves, probably indicating the realm and reach of Indian Buddhism before it sublimely merged with Bon and Tibetan influences.

Day 8: Yara - Tangge

Over the next three days, we trudged along a relentless trail that looped back southwards towards Muktinath. Descending from Yara we once again crossed the Puyon khola and ascended onto a large canyon plateau with the summit pyramid of Dhaulagiri looming over us. Soon we descended into the deep canyon of the Dhechyang khola, a major tributary of the Kali Gandaki and forded its icy waters to reach the relocated village of Dhye on its south bank. A very steep ascent followed on a hill called Sertang Danda. The trail looped endlessly through the convoluted slopes until we reached the pass Tangge la. A steep descent from the pass followed by a long traverse through sculpted switchbacks brought us to a very large plain. Traversing it, we descended once again down the fluted canyon walls of the Tangge khola valley, another tributary of the Kali Gandaki with the pretty village of Tangge ensconced at its confluence with the Yak khola on its south bank.

Day 9: Tangge - Tetang

Crossing the Tangge khola we forded the tributary Yak khola and ascended the canyon walls on to mixed terrain as it switch backed over small dry valleys. We came to the prayer flag decked Paha la with great views. A traverse followed after which we reached Paha where there was a herder`s shed in ruins.

A very long traverse followed over alpine slopes after Paha and the path could be seen way ahead. Soon we came to another view point where we could see almost the whole route we had come by all these days. Towards north were the white specs of Kore la, the green patches of Lo Manthang, Tsarang, Ghemi and even the red cliffs of Dhakmar. On our west was the road to Shyangbochen and the village of Geling. A slippery trail descended into the valley of the Narsing khola on whose banks was the village of Tetang. Taking the left fork, past another small village Chhomnang, we forded the Narsing khola to reach Tetang. All the houses had very high walls and the whole village looked like a fort. It took us a while to locate a tea house in the fading light.

Day 10: Tetang – Muktinath - Jomsom

Engulfed by scree cliffs all around with great views across the Kali Gandaki, Tetang would be the last Tibetan village on our trek. A short steep climb along the Dhingklo khola brought us to the last pass, the Gyu la. From here the prominent peaks bordering the Thorung la, the Yakwakang and Khatung Kang were seen, besides the Nilgiri, Dhaulagiri and Tukuche. To our disappointment we could also see a trail descending to the town of Muktinath. The mystical Mustang magic was over all too soon. The holes in the hills, portals to the past had been gracious to our prying presence while trying to hold on to their sepulchral secrets. As it is said, some of the best moments in life are the ones that you can’t tell anyone about.

Trekkers: Mahesh Kumaraswamy, Prakash Nuggehalli and Lakshmi Ranganathan.

Duration: 19 September to 29 September, 2012.

Summary

Trek through the Upper Mustang region of Nepal Himalaya.

14

Illusion in Mekong Headwaters and Enchantments in Highlands

A Journey through Eastern Tibet

Tamotsu (Tom) Nakamura

I returned home on 26 June 2013, from a three week 4500 km journey through eastern Tibet: West Sichuan-Qinghai. Unfortunately the first objective to explore hidden and untrodden peaks and glaciers in the remotest Mekong river headwaters was not attained, since a horse caravan could not be organised because almost all nomads had gone out for digging caterpillar fungus. In spite of my Tibetan guide Awang's two-day- effort to find nomads to help us at the 4700-4800 m high headwaters, we had to give up gathering horses and horse keepers. I shall return in August of 2014.

Nevertheless, the journey was filled with an abundant harvest.

Members and Itinerary

An elderly pair, Tom Nakamura (78) and Tadao Shintani (69) were the members supported by a capable English speaking Tibetan guide Awang (36), a Tibetan cook, Tshering(43) and two Han drivers, You hong wei (45) and Zhang qi juu (40). Two Mitsubishi Pajero vehicles were used.

7 June : Chengdu to Kangding (2500 m)
8 June : Kangding – Chetoshan pass (4298 m) – Kangding Airport (4250 m) – Tagong Grasslands – Ganzi (3400 m)
9 June : Ganzi – Manugango – Sichuan/Qinghai border pass (4700 m) – Yushu (3650 m)
10 June : Yushu, visit to Jeku monastery.
11 June : Yushu – Border with Zadoi County (4250 m) – Changlashan pass (4712 m) – Zadoi (4020 m)
12 June : Zadoi – crossed 11 high passes between 4500-4800 m – Hongse village (4700 m) which is a base for heading to Mekong Headwaters.
13 June : Unsuccessful efforts to gather horses
14 June : Hongse – access to Mekong headwaters (for tourists) – Zaqing (4260 m) – Zadoi

15 June	:	Zadoi – Yushu
16 June	:	Yushu – Three Rivers Park – Bayanharshan pass (4824 m) – Madoi (4220 m)
17 June	:	Madoi – Yellow river source-Lake Ngoring (4250 m) – Madoi
18 June	:	Madoi – panorama north face of Amnye Machen massif – Xueshan village (3780 m) – Maqin (3780 m)
19 June	:	Maqin – Qingqinshan pass (4382 m) – Gandu (4040 m) – Dari (3980 m) – Bayu (3830 m) – Longgeshan pass (4398 m) (panorama NW face of Nyaibo Yuze range) – Nyainbo Yuze Cottage (4100 m)
20 June	:	Cottage – lookout point of lake and east face of Nyainbo Yuze – Bayu – Pema (3560 m)
21 June	:	Pema – a pass at 4510 m – Border Qinghai-Sichuan (3750 m) – Seda / Serthar (3860 m) – Serthar Buddhist Institute (3800-4200 m) – Institute guest house (3925 m)
22 June	:	Guest house – Barkam (2630 m)
23 June	:	Barkam – Chengdu

Uncertain Frontiers caught in a hive of Western Development

Construction of dams and new roads with tunnels is progressing along the Dadu He river basin from Ximien to Kangding/Danba. Mountain slopes and surroundings are being mercilessly destroyed. Kanding town is developing very fast and a new airport at 4250 m near Cheto Shan pass (4298 m) has started operations. A new by-pass road has been built from the airport to Tagong monastery through beautiful grasslands pastured with yaks and sheep. Incessant destructive construction continues on the frontier of the West Sichuan Highlands and Qinghai Province. New construction and refurbishing of monasteries are also on in full swing. You would be surprised to see the construction rush of monasteries everywhere in the Kham and Amdo Tibetan territories. The glittering monastery of Tagong, unique prayer flags of Manugango lamasery, rapid expansion of historical Zogcheng monastery in Sichuan and large scale reconstruction of Jeku monastery in Qinghai heavily damaged by the earthquake in 2011 are a great attraction for travellers.

Over almost all areas where Tibetan nomads inhabit in eastern Tibet of Kham and Amdo, a drastic migration policy is forcibly being executed. New compounds known as 'New Village' with compact and modern houses are being built near towns or close to main roads. This is resulting in a rapid change of lifestyle for nomads. They stay in their new house in winter and live in a tent for pasturing in summer.

Our serious concern was the restriction on foreigners imposed by the public security bureau (PSB) since last year in the counties where fire suicide by Tibetan monks had often taken place. Fortunately, however, we were checked only one time between Ganzi and Manugango, a junction to Chola Shan pass / Dege and Yushu of Qinghai. But we had to pay 400 RMB for two cars - the reason given was excessive loads. (The officer first requested 4000 RMB.)

The First Two Objectives – Ganzi to Yushu / Zadoi

Before reaching the main objective, the Mekong Headwaters, we had two other objectives as mentioned earlier i.e. to reconnoitre the following peaks in the two areas, where reconnaissance and climbing have not yet happened. They could be future alpine paradises:

Chola Shan East massif, Sichuan

Peaks north of Manugango between a road to Yushu and Sichuan-Tibet highway (north) are totally untouched. 5200 – 5500 m rock peaks range south to north. Fortunately almost all peaks have been photographed and recorded. The main peak, Chola Shan I (6168 m), is now a popular mountain for commercial expeditions organised by Chinese companies.

Monument near Yellow River headwaters. (Tom Nakamura)

Rock peaks in the Upper Mekong near Zadoi, Qinghai

This area is hard to explain from a geographical point of view. However, if you travel from Yushu town to Zadoi in the upper Mekong, you will come across many alluring peaks though they are less than 6000 m.

(1) North of Yushu town between Yushu and border with Zadoi County: many 5400 – 5700 m impressive snow and rock peaks are viewed.

(2) Between Yushu/Zadoi county border and Dadoi town: There are two clusters of 5200 – 5400 m outstanding rock peaks. A panorama from Changlashan pass (4712 m) is breathtaking. One massif has peaks closely seen from the pass. The other massif is located closely east of Zadoi town through which the upper Mekong river flows west to east.

(3) Area surrounding Zadoi town has also many fascinating peaks between 5300 – 5800 m but yet to be surveyed. They are viewed from a road Zaqing northwards to the Mekong headwaters.

Yushu (Jyekundo) – Cross roads of early explorers

Yushu was an important crossroad of the traffic to Lhasa from the east. Famous explorers and missionaries stayed in or passed through Jyekundo towards west or east. Some of them are:-

December 1881 – January 1882: Pundit 'AK', British India

May 1889: William Rockhill, an American diplomat

November 1892 – January 1893: Miss Annie Taylor, China Inland Mission, British

May – June 1894: De Rhins and Grenard after reaching a Mekong river source, French

September 1898: Dr. Suisie C. Rijnhart, a missionary from the Netherlands

August 1900: Kolzoff, a Russian explorer

April 1918: Erich Teichman, British Consular Officer

May 1922: Brigadier George Pereira, British and Alexandra David-Neel, French

To the upper Mekong, Zadoi, and problems of 'Caterpillar Fungus'

Fine weather continued. We headed to Zadoi from Yushu on 11 June. A new airport (3800 m) 30 minutes north of Yushu town is in service.

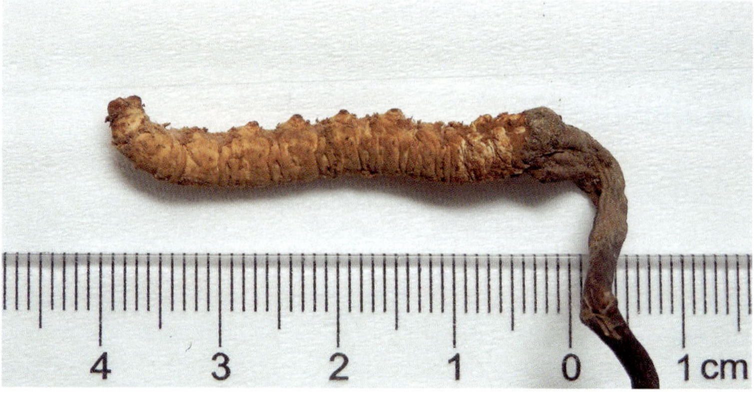

Caterpillar fungus. (Kenzo Okawa)

In an hour's drive from the airport we crossed a pass of 4504 m, passed a junction 4250 m to Nangcheng and then we reached a check post in Shizu Station at the county border of Yushu and Zadoi. A problem arose here. All travellers were checked at this check post not only by the Public Security polices but county officers and representatives of villagers. The reason of inspection is to prevent outsiders (non-residents of Zadoi County) from entering Zadoi to gather caterpillar fungus now sold at a very high price. June and July is the season of cropping. Bloody fights and disputes are not unusual to guard territory from the outsiders in eastern Tibet. Zadoi town is known as a base for exploring the source and headwaters of the Mekong river. The surrounding landscape features the beautiful upper Mekong basin but the town itself is very dirty and uncomfortable to stay. During June to July, primary schools are closed for caterpillar fungus gathering and many street

Water birds near Mekong Headwaters. (Tom Nakamura)

shops are shuttered. The same takes place in Nagchu too. My guide Awang says that inhabitants in Zadoi are unfriendly Khamba while Amdo people are friendly.

On 12 June, weather was still fine and we departed from Zadoi northwards to Hongse (4700 m), one of the four villages of Zaqing District, which is a starting point to organise a horse caravan for heading to the main objective, untrodden and unknown large glaciers and many hidden 5800 m peaks. They are recognised on the Chinese 1:100,000 topographical map and Google map as well. A road from Zadoi town crosses the Mekong to the north and directly ascends straight up to a 4750 m pass to Zaqing District. Rock peaks seen eastwards are picturesque. These are part of the Qiajaima massif that has the highest peak 5930 m in the Mekong headwaters in the north of Zaqing. This peak was climbed by a Japanese party in 2004.

We further crossed nine high passes of 4400 – 4700 m and arrived at Hongse (4700 m) late afternoon. There is only one nomad house and the Communist Party' meeting office. We stayed in a nomad family tent.

Illusion of Unknown Peaks and Glaciers

Our plan was drawn up in consultation with Masayuki Kitamura of the Tokyo University of Agriculture. Kitamura is an expert on the

Maqen Kangri II (6268 m) east face. (Kuniaki Sakai)

Maqen Kangri I (6282 m) east face. (Pedor Detjen)

Nyainbo Yuze (5300 m) north face. (Yasoyuki Takeuchi)

Mekong. He first reached the true source of the Mekong in 1994 and later accomplished the complete descent from the source to the river mouth in Vietnam. He recommended Hongse village to start a horse caravan to the destination. He advised two routes along river streams to reach the northern side of the area that we were to explore.

As soon as we arrived at Hongse, we first asked our host from the nomad family to arrange for a horse caravan. But his reaction was discouraging. As Kitamura had warned, almost all horses and nomads in the vicinity were out in the mountains digging caterpillar fungus. We visited two nomads houses far from each other in the wide open valley of the headwaters at 4700 – 4800 m hoping to find horses and horse keepers but in vain. However, fascinating landscapes of the plains, valleys, streams with remaining snow beds, 5400 – 5500 m rock and snow peaks and fauna such as water birds and deer allured and enchanted us.

The following day, 13 June was cloudy with soft sunbeams. We carried on looking for horses. We drove along the southwestern foot of the mountains, beyond which there must be the veiled peaks and glaciers of our objective. Nomads told us to come in August after caterpillar fungus season of June and July is over, as enough horses would be available and weather would be warm and stable. On 14 June, we returned to Zadoi and then to Yushu on 15 June for continuing our journey to the next stage.

When writing this article, an email came from the Netherlands. A Dutchman and an Italian reported that they had reached a new source of the Mekong in Jifi Shan and sent me a picture of mountains of our objective taken southeastwards from the Jifi Shan. It is invaluable information to unveil the hidden mountains.

Bayanharshan and Source of the Yellow River

We proceeded to the next stage: (1) Bayanharshan and Yellow river headwaters (2) Amnye Machen massif (3) Mountains in Golok, Nyainbo Yuze (4) Serthar Buddhist Institute, Sichuan.

We followed a road eastwards from Yushu on 16 June. Here too we saw a hive of development in progress. A new highway from Yushu to Xining was being constructed. The road soon crosses the upper Yangtze, Tiantong He, and continues through 4000 m high fertile grasslands towards the Bayanharshan pass (4824 m).

Bayanharshan is a name of the main peak (5262 m) of long and huge Bayanharshan mountain range stretching northwest to southeast sharing the watershed of the Yellow river (Huang He) and the Yangtze river (Chang Jiang) in Qinghai Province. But the mountains are not high.

Stone tower near Li Xian. (Tom Nakamura)

The western rim is Yagradagze Shan (5214 m) and the eastern rim is an outstanding Nyainbo Yuze (5369 m) massif of the Golok mountains.

We stayed at Madoi (4220 m), a newly developed town, for two days to visit the Lake Ngoring (4250 m) of the Yellow river headwaters. Peaks of the Bayanharshan were seen beyond the turquoise blue lake of Ngoring with water birds and a herd of Yaks.

The Bayanharshan (5267 m) was ascended by a Japanese party, members of the Japan-China Joint Yellow River Headwaters Expedition 1985. But the climbing team found a wooden turret on the top, perhaps brought by nomads.

Amnye Machen having long been Mountains of Mystery

After visit to the Yellow river headwaters, on 18 June we headed to the Amnye Machen massif. Leaving the main road to Xining we entered a road going southeast for reconnoitering the Amnye Machen. Fortunately the weather was not bad. First we saw a whole panorama of the north face, in the middle of which was the gentle but highest Maqen Kangri I (6282 m). Later we could identify all the peaks. Then we crossed a 4500 m pass and drove down a valley of Weigela in the eastern side of the massif. Peaks after peaks appeared. We passed through Xueshan village (3780 m) and lodged at a sizeable town of Maqen, a county capital.

It is well-known that Amnye Machen had long been 'Mountains of Mystery' until the mid 19th century. The first westerner to have seen the mountains was British Brigadier George Pereira. On the way from Beijing to Lhasa in 1922, he saw it from a distance and assumed its height to be some 7500 m. Inspired by Pereira's information, an explorer and plant hunter, Joseph Rock organised an expedition sponsored by the National Geographic Society in 1929 and its record was published in the National Geographic Magazine – February 1930[1]. A paragraph from that article is quoted as below:

> After dangerous, difficult months, I reached the headwaters of the 2000-mile long Yellow river and the towering, unexplored range of Amnyi Machen, twenty-eight thousand feet, or almost as high as Everest, its tallest peak lifts its snow-white head, majestic as the Matterhorn.

Joseph Rock's expedition started at Jone in Gansu Province and via the famous Labrang monastery reached present Maqen in Qinghai Province. In 1949 an American pilot, L. Clerk accessed Amnye Machen but its accurate altitude could not be measured.

It had to wait till the new China was constituted to unveil the Mountains of Mystery. In June 1960, the Beijing Geology Institute expedition made the first ascent of Maqen Kangri II (6268 m). The Chinese party surveyed and researched for two months. They identified the highest peak and other peaks as under.

Main summit – Maqen Kangri I (6282 m), II (6268 m), III (6090 m), IV (6070 m), and V (5966 m).

The Deng Xio Pini' reform and open-door policy in late 1970 led to making climbing possible. On 22 May 1981, a Japanese party from Joetsu, Niigata, made the first ascent of the main summit Maqen Kangri I (6282 m) from the eastern side via Halon valley, and a German party climbed soon after. In autumn of 1984, a Japanese party from Nagano ascended Maqen Kangri II. In 1993 a Japanese party of Tottori University climbed Peak III of Amnye Machen group (6090 m). This peak was first scaled in 1984 by the Wuhan Geology University, China.

Golok and Nyaibo Yuze – Serthar Buddhists Institute

The Golok Tibetan Autonomous Prefecture of Qinghai is a region between the mountain ranges of Amnye Machen and

1. 'Seeking the Mountains of Mystery: An Expedition on the China-Tibet Frontier to the Unexplored Amnyi Machen Range, One of Whose Peaks Rivals Everest'

Golok horseman. (Tom Nakamura)

Bayanharshan. The southern side of the Bayanharshan East forms a complicated topography with valleys that is connected to the Hengduan mountains in Sichuan. Annual average temperature is minus 4^0 C, and annual rain fall is 500 – 700 mm. The other impression of Golok is that it used to be a notorious borderland of savage and brutal Tibetans. They sometimes attacked trade caravans and even explorers. However it is now a peaceful land with fertile pastures to herd Yaks and magnificent monasteries.

On 19 June, we crossed four high passes of 4200 – 4500 m heading to Nyainbo Yuze. At 2.00 p.m. we stood at the Longgeshan pass (4398 m), the best lookout point of the northwest face of mighty rock peaks range of Nyainbo Yuze massif. It is a really breathtaking panorama. We were overwhelmed and touched by the landscape.

Nyainbo Yuze is a local Tibetan name of the Golok mountains. The massif is about 40 km south to north and 20 km east to west with deep valleys and many glacial lakes. Numerous granite rock peaks and pinnacles tower to sky ranging like the teeth of a saw. They may be compared to the rock peaks of Siguniang massif of the Qionglai mountains in Sichuan. The northern side of the main summit (5369 m) with beautiful lake, Ximen Co, is now becoming a touristic spot for Chinese visitors. However, almost the entire massif remains unfrequented in the other area, and must draw attention as a climbing paradise in future like the Qionglai mountains where climbers are rushing.

In 1989 a Japanese party from Kyoto attempted to climb the main summit (5369 m) from the lake, Ximen Co. The party set up a base camp

Serthar institute. (Tom Nakamura)

Serthar sky burial yard. (Tom Nakamura)

above the southern end of the lake and followed a route along a glacier. However, they could not stand atop the highest peak but climbed an adjacent peak a little bit lower. After the Japanese expedition, a team of the Qinghai Mountaineering Association succeeded in the first ascent of the main summit (5396 m), but their climbing route and year of the ascent are yet to be confirmed.

On 20 June, we left Nyainbo Yuze and returned way via the Serthar Buddhist Institute, Sichuan. The Institute has been opened for foreigners since March 2013. We spent a night in a guest house of the Institute at 4000 m. The institute is the largest Tibetan Buddhists Academy in China with 20,000 residents including lamas, teachers, monks, nuns and students being administrated by the government. The whole compound consisting of temples, schools and accommodations spreading over at 3800 – 4000 m in a valley are huge and spectacular. A mani-wheel symbol building is brilliantly lit up at night. A wide sky burial yard with an awful tower has been newly constructed outside the institute.

We safely arrived at Chengdu on 23 June, finishing a 4500 km long journey through Sichuan – Qinghai – Sichuan.

Summary

An exploratory 4500 km long road journey observing mountains of Sichuan-Qianghai undertaken by two senior explorers from Japan.

15

Himalayan Heat

George Mallory and Glacier Lassitude

Dr. Harvey V. Lankford

The term 'glacier lassitude' has had an endearing and enduring usage by mountaineering authors. For the first time, this article traces the origins and use of 'glacier lassitude' from the British and Himalayan mountaineering literature. The object of this work is to find and give credit to the person or persons to whom we owe this fine expression and its literary tradition thereafter. It may be a trademark expression of none other than famous Everest pioneer George Mallory.

Outdoor enthusiasts and medical providers alike are familiar with heat-related illnesses. The Himalaya seem an unlikely location for this to happen, but it does. Lassitude is a feeling of weariness, listlessness, sluggishness, lethargy, torpor, languor, or diminished energy from physical, mental, or emotional causes. John West's authoritative textbook *High Altitude Medicine and Physiology* has no direct mention of the term glacier lassitude, but this description: 'Mountaineers are familiar with the extreme intensity of solar radiation, especially on a glacier in a valley between two mountains. Here the sunlight is reflected from both sides as well as from the snow or ice on the glacier and the heat can be very oppressive despite the great altitude.'

In *The Altitude Experience*, an educational guidebook, Farris portrays lassitude as not just a physical problem but also 'negative will,' distinct from laziness that is chosen whereas lassitude is decreed. 'The best example is glacier lassitude. There is nothing in the world more able to evaporate will than trying to carry a load up a hot, sunny glacier.' The continuum leading to heat stroke is described as starting with glacier lassitude, 'the combination of dazzling light, heat, and snow seem to suck the will to move from your veins and makes walking or climbing a horrible chore.'

Thomas George Longstaff, the English physician and explorer, was the first person to climb a summit of over 7000 m, Trisul in the Indian Himalaya in 1907. His book *Mountain Sickness* recounted high altitude illness noted by western explorers over the centuries starting with

Cortes' men in 1519 on Popocatepetl in Mexico. Longstaff stated with accuracy, ''Mountain lassitude' which few can escape at altitudes of over 19,000 ft.–we refer to this height an altitude beyond which deterioration outstrips acclimatisation,' but he used the incorrect understanding of his time, 'Stagnation of the air on snowfields has long ago been suggested as a cause of mountain sickness.'

Charles Granville Bruce climbed with early Himalayan mountaineering pioneers Collie, Conway, Younghusband, Mummery and others. He was leader of the second British Expedition to Everest (1922) as well as the third (1924) until he became ill and Norton took over. In *Mountain Sickness*, Longstaff attributed the term 'mountain lassitude' to Bruce, with whom he explored in 1907. Longstaff said that earlier in 1892 when Conway and Bruce were in the Karakoram that Conway 'complains of the scorching rays of the sun when traversing the snowfields,' and Bruce 'has invented the term "mountain lassitude" by which he means "diminution in the strength of a man due to diminished atmospheric pressure".'

Alexander Kellas. Kenneth Mason in his 1955 *Abode of Snow* wrote, 'Perhaps no climber has enjoyed himself more among the Sikkim Himalaya than Dr. A.M. Kellas of Glasgow.' While Kellas was a Scottish lecturer in chemistry and not a physician, he 'Made observations on his physiological reactions to altitude, and wrote valuable accounts on these aspects.' Going over 20,000 feet many times in the Himalaya helped Kellas to understand altitude illness. The reader is directed to a review of Kellas' activities in Rodway and Mitchell's 2011 book, *Prelude to Everest.* In a 1916 reading by Kellas to The Royal Geographic Society and based on his observations of oxygen utilisation on other Himalayan peaks, he correctly predicted that, 'near the top of Mount Everest…the rate of climbing might be less than 300 feet per hour.' Kellas' larger 1920 manuscript, a masterpiece far ahead of its time, had a specific goal, 'A Consideration of the Possibility of Ascending Mount Everest.' It lay dormant at the Alpine Club in London, given to them in 1920 but finally published in 2001 for the first time by John West in *High Altitude Medicine & Biology.* This larger Kellas work includes a section on mountain lassitude. Written in 1920, Kellas the expert did not specifically mention glacier lassitude because he died near the Tibetan village of Kampa Dzong on the approach march of the 1921 First British Everest Reconnaissance Expedition weeks before they entered the Rongbuk glacier area and before George Mallory ever uttered the words.

The next two quotations are from Kellas' section on physiological difficulties and mountain sickness. Kellas appropriately discarded other theories such as ionizing, electrical, or even psychic origins. Instead, sounding much like John West's future textbook of 2007, he emphasised that, 'It is generally assumed that snow, especially melting snow under a hot sun, is far more likely to cause trouble than rock, and a ravine or gully with stagnant air is more likely to produce mountain sickness than an open ridge ... it is far more likely that the climber is influenced by the intense reverberation of the sun's rays from the snow.' In his description of 'General malaise: lassitude' there is an explanation that, 'Mountain sickness varies within certain narrow areas....passes of about the same height in the same region differ greatly as regards to mountain sickness....it has been generally assumed that one is more liable to be attacked when climbing in gullies and on snow, than on open ridges and rock.' This sounds like glacier lassitude. Kellas and other writers that he referenced from as early as 1783 already knew of the problem - it just awaited a catchy name. This was what George Mallory found out later, and verbalised, in 1921-1924.

George Leigh-Mallory

It is ironic that on the first British Everest Expedition of 1921 medical observer Alexander Kellas died while mountaineer George Mallory described glacier lassitude, whereas on the third British Everest Expedition of 1924 it was the other way around: the mountaineer George Mallory died while medical officer RWG Hingston officially reported glacier lassitude.

George Mallory, in his 1921 letters from Everest base camp complained that, 'I found it pretty hot on the glacier...I must confess to a degree of tiredness after the glacier work which I have never quite reached in the Alps.' In Bruce's *Everest Reconnaissance: the first expedition of 1921,* Mallory's section included the description of glacier and lassitude on the Tibetan Main Rongbuk glacier: 'It was the glacier that had knocked me out, not the hard work alone but some malignant quality in the atmosphere, which I can neither describe nor explain; and in crossing a glacier during the day I always afterwards observed the same effect; I might feel as fit and fresh as I could wish on the moraine at the side but only once succeeded in crossing a glacier without feeling a despairing lassitude.' Mallory repeated it in chapter 16 on weather and condition of snow, 'More than once, though the sun was not shining, in crossing a glacier late in the day I was reduced from a state of alert activity to one of heavy lassitude.'

In Mallory's 1924 letters, published posthumously in Norton's book on that fatal expedition, he depicted 'the Trough' of the East Rongbuk glacier like this, 'Irvine suffered very much and I somewhat from the complaint known as glacier lassitude – a mysterious complaint, but I am pretty certain in his case that the sun and the dazzling light reflected from the new snow had something to do with the trouble.'

The use of 'lassitude' involving explorers before 1921 such as Bruce, Conway, Longstaff, Morshead, Whymper, Wollaston, Workman, Younghusband and others revealed that the word was used in the traditional sense of fatigue in the mountains, but not the glacier-specific usage as in Mallory's 'glacier lassitude.' Mallory's use of 'known as' suggests he might not have been the first to say glacier lassitude in 1924, but he might have invented it earlier during that trip, or in 1921 on the main Rongbuk glacier, or in 1922 on the East Rongbuk glacier. Many others seem to give credit to Mallory as discussed below.

Norton , Odell, and Younghusband

Edward 'Teddy' Norton, the leader after Bruce became ill, reported in *The Fight for Everest 1924*, that high on the East Rongbuk glacier at Camp III (21,300 feet) that, 'Somervell – had more the matter with him than the early morning lassitude; he was in fact suffering rather severely from the sunstroke he had contracted in the windless glare of the 'trough' the day before.' The 'trough' or medial moraine was part of the problem. The sun was strong there in the thin air and the location is closer to the equator than, for example, the Great Pyramid in Egypt. Looming tall seracs and other ice formations surrounded the team on both sides. Scientists refer to 'albedo' - the extent to which something reflects light from the sun. Fresh deep snow over a featureless landscape may have an albedo as high as 90%. Glacial hard ice and/or an irregular surface may be less, however, the fact that the Trough climbers were surrounded on both sides suggests that they were thermally ambushed in a glacial slot canyon, thus experiencing not high-altitude lassitude alone, nor heat lassitude alone, but the two of them combined into the Himalayan super grand total of glacier lassitude.

By the time of the third British Everest Expedition, scientific understanding had improved. Noel Odell, in his chapter on Geology and Glaciology in T*he Fight for Everest 1924*, explained with scientific precision what happened on the East Rongbuk glacier in 1924 and the previous expedition of 1922, and similarly on the main Rongbuk glacier in 1921, 'In spite of the low mean annual temperature from which this region suffers, its situation in such low latitude (Everest is 28° N) causes it to experience the effects of a very hot sun [and] a striking example of the effects of a tropical sun is the way in which eastward and westward flowing glaciers frequently have their southern sides melted back into steep ice cliffs, a case in point being the tributary to the East Rongbuk at Camp II...this action is aided by re-radiation of heat from rock-walls enclosing the glacier.' Odell continued with his description of the medial moraine, '*The Trough* – This interesting and possibly unique feature that provided a natural causeway for nearly two miles of the way up and down the East Rongbuk glacier between Camps II and III, will ever be remembered as a fairy scene, of the greatest beauty and highest artistry... exhilarating in the extreme, except at the hot hours of the day in late May and June, when the close stagnant air within the depression of the Trough was apt to produce a certain lassitude and disregard for its remarkable beauty!'

So now we have these facts; we have arrived back to the three expeditions of the early 1920s, we have the glaciers, their location and description, the source of the heat, and the presence of mountaineers. In his 1926 trilogy about the three expeditions, *The Epic of Mount Everest,* the first Mount Everest Committee Chairman and Great Gamer Francis Younghusband wrote that the 'party felt a peculiar kind of lassitude which evaporated all the energy in them. It was what afterwards became known as "glacier lassitude",' Note his choice of the word 'afterwards.'

Richard William George Hingston served as the Medical Officer for the third British expedition to Everest, surprisingly reaching Camp IV (27,400 feet) and contributing to Norton's book *The Fight for Everest 1924.* Identical paragraphs occurred in that book and in Hingston's 'Physiological Difficulties in the Ascent of Mount Everest' appearing in *The Geographical Journal* in 1925, with a section titled 'Glacier Lassitude.'

Hingstons' first paragraph stood squarely on the shoulders of Longstaff, Kellas, Mallory, Norton, and Odell. A second paragraph, however, became problematic when he blamed 'saturated atmosphere.' A 1925 *Lancet* medical journal article discounted humidity as possibly contributing to glacier lassitude. Leonard Hill and Argyll Campbell at the London National Institute of Medical Research discussed the physiological problems of the Everest expeditions and agreed that Hingston 'clearly demonstrates that great lassitude was produced by overheating the body whilst at a high altitude.' Hill and Campbell studied the effect of exercising while breathing air saturated with moisture and at low oxygen tension on six test subjects, simulating 20,000 feet in the laboratory. They concluded that while 'Hingston refers to the air being saturated with moisture in the case of glacier lassitude... this factor is almost negligible,' and finally, 'it may be concluded that "glacier lassitude" was due to the combined effects – of overheating of the body and of breathing oxygen at low tension.' The humidity theory died in the scientific literature, but the term glacier lassitude was to gain immortality in the mountaineering literature.

Mountaineering Literature and Glacier Lassitude

The British mountaineer and prodigious romantic writer Frank Smythe wasted no time theatrically plugging the phrase into his literary hobby. He was a frequent embellisher, most famously known for his hallucinatory companion on Everest. In *The Kangchenjunga Adventure* and going badly on a glacier in 1930, he rightfully attributed the term glacier lassitude to the 1922 and 1924 Everest teams, 'Members of the Everest expeditions remarked the same thing in the trough of the East Rongbuk glacier.'

His artistic flare with glacier lassitude shows in three examples: from *The Kangchenjunga Adventure* in 1930, 'The sun poured down upon us a fierce heat in which the snow became more abominable every hour [and] once again we experienced the energy sapping effects of glacier lassitude,' from *Kamet Conquered* in 1932, 'Our bodies and minds became the unwilling slaves to that insidious drug, glacier lassitude [and] crossing the glacier...we felt like flies in the focus of a burning glass [for] the sun shone with relentless vigour upon our backs and our pace was funereal,' and from *Camp Six* describing Everest in 1933, 'On the snow-covered glacier there was such an airless blinding glare that our energy drained away like liquid through a funnel.'

In his 1955 condensation of multiple expeditions entitled *Men Against Everest*, famous British explorer Eric Shipton described the scene on the north side of Everest in 1921, 'As they climbed slowly up the glacier they were oppressed by a curious lassitude which robbed them of much of their strength. It was what has since become known as "glacier lassitude".' Note his choice of the words 'what has since become' appears to date the term to 1921.

Matt Dickinson's *The Death Zone* gives this credit, '"Glacier lassitude"- the disease so brilliantly introduced to the world in the book, *The Ascent of Rum Doodle*' however, it was actually only exaggerated there. W. E. Bowman's *Rum Doodle* is a 1956 parody of a bumbling, thoroughly British Himalayan expedition, supposedly inspired by the 1934 & 1936 expeditions rendered more classically in Shipton's *Nanda Devi* and/or HW Tilman's *The Ascent of Nanda Devi.* At most, *Rum Doodle* gets credit for bringing many lassitudinal words to the armchair traveller. Bowman mentions London lassitude three times, heat lassitude once, valley lassitude once, plain lassitude once, sleeping bag lassitude once, altitude lassitude once, base camp lassitude once, and the vaunted glacier lassitude twice.

In conclusion, Longstaff, Kellas and others knew the symptoms of mountain lassitude, high-altitude lassitude, calm-air lassitude, and narrow-space lassitude, but never uttered the famous term 'glacier lassitude.' Mallory first used the two words glacier and lassitude in the same sentence in 1921. Mallory combined them into 'glacier lassitude' before he and Irvine disappeared on Everest in 1924. Norton and Odell added their descriptions in 1924. Hingston authored his medical reports in 1924 and 1925. Researchers Hill & Campbell clarified the phenomenon further in 1925. Younghusband, Smythe, Shipton, and others gave credit to the early pioneers, especially Mallory, and made it a literary tradition thereafter. There may be additional sources or private papers, especially prior to 1924, not included here. Other researchers are invited to help analyse the term glacier lassitude, and perhaps give credit to the words of George Mallory, to paraphrase his famous Everest phrase, 'because they are there.'

Summary

The coining of the term 'Glacier Lassitude' by Mallory and further use by other mountain writers thus establishing the term in mountain literature

EXPEDITIONS AND NOTES

1 Mountain Festivals of the World

Bernadette McDonald

01 March 1988. It was my first day of work as Director of the Banff Mountain Film Festival. My boss called me in to his office to discuss the parameters of my job and, almost as an aside, mentioned that I should ensure that my passport was current because I would need to go to Italy in May to attend the Trento Mountain Film Festival. I retained my composure, nodded (without grinning) and said, 'of course, I'll get on that'. I tripped back to my office, grabbed the world atlas – (remember, this was 1988!) and looked for Trento. Northern Italy – fly to Milano – train to Verona – change trains, turn left and head into the mountains. Yippee!

Aside from Banff, which was my home festival and which I had attended since 1977, Trento was the first mountain festival I had the occasion to visit. It wasn't the last. Over the twenty years that I directed the Banff festival, and in the years since then when I have continued to attend festivals, either to jury films or give lectures, I have been privileged to be part of this global mountain community.

I remember one jury dinner at the Kendal festival in Britain's Lake District. American filmmaker David Breashears stood up and proposed a toast. We expected a toast to the jury, perhaps to the filmmakers whose work we were there to celebrate. But no, his toast was to the festival organisers. David explained in just a few sentences how important mountain festivals were to the mountain community. Organised sports have competitive games where enthusiasts gather; artists and musicians have concerts and openings where their community gathers. But climbers go into the mountains to practice their passion: there is no 'gathering'. Climbing takes place in relative solitude. But there *is* a mountain community – a group of people around the world who share those alpine values, those vertical experiences, that passion for going up. As he raised his glass to the festival team he thanked them for creating events where that community of loners could meet, tell stories, celebrate successes, mourn their friends who remained in the mountains and, above all, share their common passion.

It's hard to know where to begin on this whirlwind festival tour, but perhaps it is with an organisation whose birth I was happy to be part of. The International Alliance of Mountain Films was the brainchild of just nine festival directors in 2000. We all knew each other and we met when we could, sharing information, contacts, ideas and crying on each others' shoulders about funding, technology and marketing. But in 2000 we decided to form a proper association, with common goals, common standards and a sense of cooperation. Our first meeting was at the Museo della Montagna in Torino, Italy, under the leadership of its director, Aldo Audisio. The founding member festivals were: Trento, Italy; Autrans, France; Graz, Austria; Cervinia, Italy; Lugano, Switzerland; Torello, Spain, and the Museo in Torino. It's amazing to look at the Alliance now, with its 21 members representing 16 countries in four continents.

You might ask where are these festivals that seem to be multiplying like rabbits? What are their characters? What differentiates them and what do they have in common? Each one of these mountain festivals has its own character – its special flair and tradition. But the common values are a love for mountain landscapes and culture, an understanding of issues that are important to mountain communities and a common passion for alpinism. In other words, each festival is an ambassador for mountain culture.

Trento – that first mountain festival that I visited – is the oldest. Begun in 1952, this is a distinctly classy festival. You can wander the historic streets and alleyways of Trento, with architectural jewels around every corner, bubbly Prosecco in every bar, mountaineering legends sipping espresso in the sidewalk cafés, and always, that heady aroma of spring. Trento is in May, and May in the Trentino valley is magical. The festival is massive: more than a week of lectures and meetings, film screenings and book events, press conferences and coveted awards. There is an air of history and tradition that you don't find at any other festival. Trento is simply Trento.

My first visit there was overwhelmingly intimidating. Just two months into the job and I was thrust into the granddaddy of them all. I survived, met a lot of important mountain personalities, saw more than 80 films and returned more than 20 times. Trento is now almost like a second home to me. Although it has evolved over the years, Trento retains its solid, indestructible character.

The second festival that I visited on this global tour was Autrans, France. Invited to be on their jury in 1989, I was surprised to learn that the village of Autrans, set in a high rolling plateau above Grenoble, was home to the x-country events for the 1968 winter Olympics. The festival has a more youthful atmosphere than Trento and is very, very French. French and Swiss filmmakers swarm to the event, creating an artsy, professional atmosphere amidst the rural setting and strong aroma of Gitanes. Autrans is also the original home of the famous Piolet d'Or – the Academy Awards of alpinism. Although the Piolet now takes place in Chamonix, the first few years were in Autrans, adding a touch of elitism as the world's finest alpinists gathered to await the judgment of the jury and their peers. How can I ever forget the first time I met Tomaž Humar as he stormed around the festival, or Lynn Hill, that tiny jewel of a woman in her even tinier black miniskirt? Even the French were wowed!

One of the most alpine locations for a mountain film festival is Les Diablerets, Switzerland. The picture-perfect village is 1160 m high on the north side of the glacier-covered Les Diablerets massif in the Bernese Alps. Founded in 1969, the festival takes place in September when the valley has emptied itself of busloads of tourists and replaced them with cows. Lots and lots of gargantuan cows, trudging up to the alpine meadows to fatten up during the day, and waddling back down to the security of their barns at night. It's quite a show. Almost as good as the films! The festival atmosphere is permeated with the guiding tradition in Switzerland. One of the highlights of the festival is the fetes des guides – an alfresco culinary extravaganza that focuses on one simple dish - raclette.

Lugano, Switzerland hosts one the most elegant of all the mountain festivals, although it would be a tight race with Graz, Austria. More of that later! Lugano's festival is different, for it is a 'festival dei festivals' – an event that showcases the best films from the other festivals. Because it is not competitive, it lacks the sort of 'frisson' that you find at competitive festivals. You can't replicate that air of nervousness and excitement. But the well-moneyed streets of Lugano and the exquisite backdrop of Switzerland's Ticino region give it a solid, self confident air.

Graz, Austria, has all of that and more. The aristocratic city of Graz boasts a long and complex history as a crossroads between the Germanic countries and the Balkan regions. Its more recent history is also controversial, for it was a favorite haunt of Hitler. With its grand houses and public buildings, spacious squares, mountainous backdrop, substantial menus and spectacular local wines, (particularly the Grüner Veltliners), Graz is well equipped to host a mountain festival. Unlike most others, the Graz festival is privately owned. Robert Schauer, one of Austria's most renowned alpinists (and that's saying a lot) is the father, mother, and patron of the Graz festival and his personality informs every aspect of it.

A much younger festival, but still one of the founders of the Alliance, was the Cervinia International Film Festival. Founded in 1998 at the base of the Italian side of the Matterhorn, this festival's original aim was to select and reward films that had already been awarded by others festivals - another 'festival of festivals' kind of approach! Since then, this festival has evolved into a competitive event as well as publishing and other special events. It's hard to compete with the location. Of course the Matterhorn dominates the view, but there are plenty of other 4000 m peaks nearby, providing lots of temptations for festival goers. I have to admit to sneaking out on more than one occasion for some nice long days in the mountains.

There are many young festivals starting up, including one in Spain's Basque region. Founded in 2008, the MENDI festival first took place in Vittoria, Spain but then moved to Bilbao for practical reasons. I returned to the festival for the second time in December 2013 and can say that this festival has extremely clear mountaineering roots. That shouldn't be a surprise, since Basque country has produced more than its share of world-class alpinists over the years. Climbing is an incredibly important part of Basque culture and you can feel it at the MENDI festival: the choice of films, the speakers, and the audience. Alpinism permeates all of the mountain cultural activities of the festival, which is young but also ambitious. It plans to develop a year-round mountain cultural program in Basque country.

Another festival with a strong mountaineering heritage is in Domžale, Slovenia. Begun in 2007, this is the brainchild of one of Slovenia's best climbers, Silvo Karo. I was at the first festival organised by Silvo and still remember the impressive lineup of speakers: Catherine Destivelle, Doug Scott, Robert Schauer, and of course the cream of the Slovenian climbers – some of the best in the world. This is an early winter festival that now takes place in the country's capital, Ljubljana, and looks like it will have a brilliant future.

A much different festival exists in Dundee, Scotland. Begun in 1983, the UK's longest running mountain film festival is non-competitive and features the Banff World Tour program together with other independent mountain films and speakers from around the world. Held the last weekend of November, Dundee has a unique atmosphere. It is wildly popular with the local population, many of whom are 'walkers' and 'Munroers'. Dundee is a 'gentle' festival – less adrenalin than some festivals that focus on climbing – on adventure – on performance. The Dundee festival treads gently on the landscape and is a welcome change.

Another important British mountain film festival takes place in Kendal, the heart of the English Lake District and the birth place of British mountaineering. The festival is packed with climbers who come from around the country for the films and lectures (I have never seen a festival with more lectures) and, if they're lucky – a nip out into the hills for a day of climbing. The fall weather is fickle but when it's good, it's very, very good. Luckily for the festival it's mostly bad, which means the climbers are indoors, watching films. Perhaps the defining attraction of Kendal is the socialising. Dare I say it – the pubs? Even the festival promotes itself as 'THE main social event for outdoor people in the UK'. They're not exaggerating. This is a tribal gathering, one of so many around the world, where the mountain community gathers and renews itself.

Most mountain festivals' characters are defined by the landscapes in which they are found. That is certainly true for Poprad, Slovakia. Poprad is a small town tucked beneath, and dominated by, the Tatras mountains which span the Slovakia, Polish border. I was privileged to be at the very first festival in 1993, plus several more since then. What a pleasure to watch Poprad find its character and its public, which is now huge. The October festival offers a symbolical 'window on the world' to its audience, and it strives to awaken and promote an interest in protecting the environment. The director, Maria Hamarova, and her husband Peter Hamor – Slovakia's top Himalayan climber – lead the festival with authority and passion and as a result, it is a real player in the European scene.

Just on the other side of the Tatras is the 10-year-old, Zakopane mountain film festival. Zakopane is the mountain stronghold of Poland. It's where this nation trained its climbers, both summer and winter. It's where Poland's incredible Himalayan dominance was formed. It's where Poland's guiding culture was born. Zakopane is a special place, with its characteristic wooden architecture, its unique mountain cuisine, and always....always...the Tatras, towering over the village.

The festival features films on mountaineering, adventure, culture, nature, and social and human rights issues. But there's much more: exhibitions, lectures, discussions, interviews, climbing competitions and mountain tours with guides clad in traditional costumes and accompanied by local musicians. This festival's strong mountaineering background is tempered by the special 'Highlander' tradition of the Zakopane valley: woodcutters, cheese makers, shepherds and free mountain spirits! This is a young festival with an exciting future. I've only been once, but I am determined to return!

Not too far away, but with a completely different atmosphere, is the film festival in Krakow. Krakow – beautiful Krakow. One of the most exquisite cities in the world – almost a distraction for this festival, which takes place in December. But obviously not, since it attracts huge audiences of mountain enthusiasts. I will never forget looking out at the auditorium, just before starting my lecture at the Krakow festival – 1200 people, waiting. This is probably the largest mountain festival in mountain-obsessed Poland.

Further north is Ladek, a Polish spa town not far from the Czech border. At least it's Polish now. Poor Ladek has flipped and flopped from Austria to Poland to Germany to Poland. For the moment it's in Poland. And although the town is usually frequented by 'spa-people', during one week in September, the streets fill up with a livelier crowd. Really quite lively! Young. Active. Full of life and energy and ready to laugh, argue and dance. They fill the temporary tent structures with hundreds and hundreds of people to watch adventure films and more than anything, listen to the incredible line-up of speakers. I've only been to Ladek once, but I was blown away by the quality of lectures. The programming wasn't just about adventure and adrenalin, but about real exploration, real issues, real history and real current events. The feeling at Ladek was of a knowledgeable festival, both from its director and its audience.

Not far away across the border in the Czech Republic is the Teplice nad Metují Mountain festival. This is a climber's town, set amongst the fantastical sandstone towers that dominate this part of the country. Begun in 1980, one of the goals of the festival is to promote mountain cinematography as well as to show – mainly to the young– a good way to spend their leisure time. It's quite curious, because one of the strangest experiences I've ever had at a festival was in Teplice. As I wandered around the village I noticed a couple of people on crutches. Then a few more! Finally I realised that I had seen at least 20 or younger, healthy-looking people hobbling around on crutches, obviously in pain from their injuries. There must be a huge hospital nearby, I thought. This must be some kind of rehabilitation centre. I was wrong. They were climbers – victims of the vicious nature of climbing on the Teplice sandstone towers. It wasn't until a day or two later that I found out just how dangerous they were - beautiful to look at, but difficult to protect oneself. If I could use one word to describe the Teplice festival, it would be 'hardcore'.

There are many more. Vancouver and Mussoorie. Telluride and Taos. Passy and Tegernsee. Kathmandu and Ushuaia. Big festivals. Small festivals. Cult festivals. Film festivals and book festivals. And at the end of this whirlwind world tour of mountain festivals is Banff. My home festival, and the one that I directed for 20 years. The Banff Mountain Festival is situated in the heart of the Canadian Rockies. It began in 1976 as a one-day event and has grown into a nine-day festival followed each year by the World Tour with over 530 screenings in about 30 countries on all seven continents. The Festival combines film and book competitions with photography and art exhibitions, a trade show, craft fair, speakers, panel discussions and writing and filmmaking educational programs. With more than 10,000 audience members in Banff and an additional 200,000+ on the world tour, it is surely the biggest single impact on mountain culture in the world. Many of those tour screenings are slowly but surely evolving into stand-alone festivals with their own distinct characters.

Banff's roots are mountaineering roots, still a strong component of the festival. But it has branched out to include environmental, sport and cultural themes. Like so many other festivals, Banff is a meeting of the tribe, a crossroads for the global mountain community that migrates each November to the little town in the shadow of Tunnel Mountain where they meet their friends, are inspired, learn new skills, and renew passion for the mountains. And that's what these mountain festivals are all about.

Summary

A kaleidoscopic view of different mountain festivals organised around the world from the author's perspective and her personal involvement with them.

2 A fortnight in Garhwal

Geoffrey Cohen

Not long ago my good friend Richard (Dick) Isherwood described in the *Alpine Journal* some climbs we did together in the Himalaya in 1980. However he omitted to mention what I thought had been possibly the most successful of our campaigns. When I asked him why, he said that as they were not new routes he didn't think they were worth describing. Now very sadly, at too young an age, Dick has gone off to climb in another world. I cannot write as engagingly and wittily as he always did, but in remembrance of a good friend here is a short account of a memorable two weeks.[1]

Richard (Dick) Isherwood. (Geoffrey Cohen)

1. See 'In Memoriam' for obituary on Dick Isherwood. - Ed.

We met up in Joshimath on 21 June 1980. Dick immediately constructed a shopping list and in the evening we went to the bazaar and purchased a generous quantity of fresh vegetables – cabbages, peppers, aubergines, tomatoes, cucumbers, onions, and of course garlic which Dick loved inordinately (he once used a surplus of several pounds of garlic to make a garlic soup!). This was supplemented by rice plus tins of butter, cheese, tuna, bacon rashers, chicken curry and a 'mystery' tin. Next morning Dick persuaded two Nepali porters to come with us to the Valley of Flowers at the princely rate of Rs 20/- per day. At that time, and possibly still, there was such poverty in western Nepal that there were always Nepalis available for work on the Indian side of the border. Our lads were sent off to get themselves shoes and blankets while we bought bus tickets for the hour's ride down to Govind Ghat, where the trek to the Sikh holy temple at Hemkund starts. After a wet night at Ghangria we walked through the Valley of Flowers, where Dick the naturalist identified for me the beautiful yellow lilies and violets, blue cranesbills, purple pea, dwarf orchids, primulas, red, orange and yellow potentillas and many others. Base camp (one two-man Vango tent) was set up at about 3960 m at the far end of the valley and our two brave lads sent back to warmer nights below.

Rataban (6166 m) northwest face. (Geoffrey Cohen)

After a day's reconnaissance we set off to attempt the northwest face of Rataban (6166 m), the most accessible and reasonable objective given our limited time. Our spare food was stashed in a small cave. We took with us two ropes, three ice screws and a few small wires. As afternoon wore on we found ourselves floundering in thick mist on a glacier shelf at about 4875 m listening to the sounds of avalanches coming off the crags above us. We reckoned we could camp far enough from the cliffs to be fairly safe and hope for better visibility in the morning. As it turned out the weather remained poor but we managed next day to weave around a variety of crevasses and ice cliffs to reach the col below Rataban, where in cold, wet and windy weather we got the tent up and crawled inside. Later in the afternoon Dick went outside and watched avalanches coming off the slopes of Rataban, one of which crossed our tracks and would have obliterated us. The following day was foul too and my memory is of snuggling down all day with our books - Dick read the recently published *Snow Leopard* by Peter Mathiessen and I had a Hindi grammar book; luckily he was very tolerant of me muttering my exercises.

On 28 June the sun began to emerge so we climbed a short way to a snowy bump northwest of the col, enjoying tremendous views of Chaukhamba, Nilkanth, Mana, Kamet and nearby Nilgiri Parvat; we allowed this day for some of the fresh snow to clear. Though it snowed again in the night and was cloudy at 5.00 a.m. next morning, after discussion we decided to attempt the peak inspite of the ominous weather. We were away by 7.00 a.m., crossing avalanche debris to reach a spur on the right of the northwest face. A little way up this we found a fixed rope; although apparently in quite good condition much of it was buried in snow and thus not very useful. We climbed eight pitches of mixed ground at about Scottish grade III, during which the cloud thinned and some patches of blue sky appeared. Either

Dick Isherwood on the northwest face of Rataban. (Geoffrey Cohen)

side of our spur the ground plunged down in steep mixed faces of impressive depth. At noon we reached a place a few hundred feet above the last rocks, where it seemed safe to unrope. Above us was a snow ridge at an angle that did not seem too challenging, so we left our sacks and just took one rope and a bite to eat. Dick stormed off up the snow at a terrific pace and I panted along in his steps. An hour and a half later, after a passing a few icy steepenings, we rested in the cloud on top with nothing to view. Dick announced with satisfaction that it was his tenth 6000 m peak, but after waiting in vain for a clearing we started down. For a couple of hours on the descent the weather deteriorated to a cold, wet afternoon, but the evening brought another clearing. Abseils were slow - we had only one stitch plate between us, as Dick had not yet acquired such up-to-date technology, so he hauled mine up for each abseil! But by dint of down-climbing where we could, we got back to camp by 7.30 p.m.

Nilgiri Parvat (6474 m) northwest face. (Geoffrey Cohen)

Next day we got down off the glacier and enjoyed a memorable lunch drying everything that had got extremely wet during the bad weather on the col, only to get more showers in the afternoon and evening at our base camp. We then took three days food to cross the Khuliagarvia col and make a lightning attempt on Nilgiri Parvat (6474 m). The way up to the pass (about 5180 m) took us alongside a beautiful waterfall and then a long grassy ridge, strewn in its lower reaches with the most gorgeous flowers. On the far side there were gentle snow slopes leading down to the desolate moraines of the Khuliagarvia glacier. We had a real sense of exploration penetrating this unfrequented place. Cloud and icefalls initially prevented us from seeing more than a brief and partial view of the north side of our peak, but having made a second camp at about 5490 m on a snow shelf we finally got a clearing on the evening of 2 July. The face above, entirely snow at a not very steep angle, looked long but eminently feasible; all we needed was a day of good weather, but we had no spare days so there was only one throw of the dice, and in the monsoon our chances were doubtful.

We were lucky and got a reasonably fine day with only thin cloud around the tops. As usual, Dick led nearly the whole way making steps in the snow with inexhaustible energy. Here and there we tried to move fast where there was some risk of serac fall, and we needed a little bit of traversing to get around some ice cliffs; otherwise there were no serious obstacles and we only needed to rope up for about three pitches. On the way up we had excellent views north towards Mana and Kamet and, much nearer, the elegant Mandir Parvat, but from the summit, views were limited by cloud. We were back in our tent after only ten hours, a fairly short summit day by Himalayan standards, feeling mighty pleased to have got this peak in only four days after Rataban.

On 4 July we descended the Khuliagarvia glacier, climbed back up to the col and back down to the Valley of Flowers. On this last descent my legs began to feel so weak that I couldn't trust them to support me, doubtless the consequence of the last few strenuous days, and when I tried to run down the last few hundred feet above base camp they crumpled completely. Dick however remained strong enough to climb back up to our stash and retrieve pack frames, walking shoes, and cooking pots. We descended next morning with 60 pound packs as far as Ghangaria, where we managed to negotiate a horse and man to take our loads back down to the road. We just made the last bus and reached Joshimath in a torrential downpour, where we gorged on coffees, samosas and burfi.

There are many styles of Himalayan climbing. Nowadays, with the growth of commercial agencies that do the local organisation, many expeditions have extensive support even for quite straightforward peaks around the 6000 m mark. Looking back at this short trip, it was

Dick Isherwood climbing northwest face of Nilgiri Parvat. (Geoffrey Cohen)

satisfying to have climbed two quite easy but respectable 6000 m peaks from different valleys in a fortnight, with little *bandobast* or expense and only a day and a half of support from two porters. Success was only feasible because our peaks were fairly accessible and thanks to Dick's huge energies. The Himalaya being so vast one hopes there will always be opportunities for small parties to be nimble and climb 'under the radar'.

Summary

An account of an early lightweight expedition to the Garhwal Himalaya by two British climbers in 1980, proving that 'small is beautiful'. They climbed Rataban (6166 m) and Nilgiri Parvat (6474 m) within a short span of four days between two summits.

3 The First Ascent of Kuchela Dhura

Dhruv Joshi

The idea of attempting Kuchela Dhura (6250 m) came when we were on the upper slopes of Changuch in 2011[1]. The north face of the peak was accessible through Lawan valley near Naspanpatti, an hour's walk before the snout of Lawan glacier. Kuchela Dhura's summit hump offers an almost three km long and sharp ridge leading towards the north face of Nanda Kot in the southwest direction. Its southern slope sharply falls into the Shalang valley. The cause of it being unclimbed is probably the attraction of other majestic peaks in this valley like Nanda Kot, Changuch and Nanda Devi East. We went in the pre-monsoon of 2012 and followed the ridge from the Nanda Kot side but only managed to reach 6206 m before we were stopped by a 'V' shaped notch.

In 2013 I opted for the post monsoon season. Earlier this year in June, as I was sitting at the end of Lawan glacier after successfully crossing

North Face and the icefall of Kuchela Dhura (6250 m). (Dhruv Joshi)

1. *HJ* Vol. 67, p. 72.

the Traill's pass with two of my climbing associates, looking at the cloud covered Kuchela Dhura, I whispered – 'We have to do it this year!' I could see only the icefall on the north face of the peak, but I had an intuition that the peak had shown us the route. I decided to climb the peak through the icefall. I was to take a few climbing members from last year's Kuchela Dhura attempt as I again planned to do the job without any high altitude supporters, but at the last minute only one participant was selected – Wallambok Lyngdoh whom I appointed deputy leader of the expedition. The new members were Vineet Kumar Saini, Chitramohan Singh Chauhan, C. Zormsanga, Banshngainlang Nongkynrih, Vijay Rautela,Ram Prasad Lodha, Karan Kumar Shandilya and our medical officer Dr. Anand Vaidya. We left Delhi on 18 August for Munsyari.

Trek to Lawan glacier and base camp

Lawan is the most accessible glacier which lies on the outer ring of the Nanda Devi Sanctuary in the Kumaun Himalaya. Lawan gad emerges from the glacier and joins Gori ganga which forms the main tributary of river Kali which is the natural border between India and Nepal. We started our trek on 20 August towards the base of the peak. It was around 70 km from Munsyari and would take around four to five days. Landslides, busted trails and wiped-out bridges were the outcome of the 16 June disaster in Uttarakhand state. As we started from Dhaba (roadhead) where jeeps dropped us, we realised that much had changed in a very short period.

Our first day's halt Lilam was not so far from the present roadhead Dhaba and it normally took around two and half hours. This time, the landslide had changed the trail and after an hour's walk we had to fix the ropes for porters as well as for members. The following day we started a bit early for Bugdiyar, 16 km from Lilam. At this moment we were on an old reopened pathway to Johar valley via Minsingh Top, which had been closed a few decades ago as the new route was along with the Gori ganga and took less time to travel the same distance. We reach Bugdiyar at 6.00 p.m., and soon heavy rain showers welcomed us. Next day we started early to reach Martoli village before the dusk. After we left Martoli village, a temporary single log bridge before Lawan village became a time consuming obstacle so we had to camp out in the vicinity of Ratagangal. We arrived at the desired location at 10.00 a.m. and established base camp. It was a pleasing alpine grazing land at an elevation of 4173 m just opposite to Naspanpatti en route to Nanda Devi East base camp on the true right bank of Lawan gad.

Base camp to advance base camp

The location of the advance base camp was in the southern direction from the base camp. The route began with a walk on a gradual grassy slope towards southeast followed by a climb then a traverse of around 400 m west led to the lower part of the glacier. After a final walk on a gradual glacial slope towards south we reached the site of our advance base camp at 4922 m. It was the same location we had reached the previous year for a recce during pre monsoon and had turned back after watching continuous avalanches from the icefall above it.

Our team made four consecutive load ferries to ABC before occupying it. On 29 August rain delayed our early start. Wallambok, Vijay, Chitramohan, Vineet, our cook Manoj and I arrived at ABC at 06.30 p.m., while C. Zoramsanga, Ram, Ban, Karan and our kitchen boy accomplished a load ferry. Rock fall and heavy rains continued through the night. Next morning we were blessed by the astonishing vision of Nanda Devi East peeping out through the clouds.

Advance base camp to summit camp

On 30 August, the first section and some part of the second section of the icefall was opened. As few ropes were fixed earlier, we moved fast. That day we fixed only 300 m of rope, and returned to ABC at 2.30 p.m.

Bad weather converted the next day into a rest day for us. On 2 September we headed for final route opening and load ferry. Karan and Ban assisted us till dump point before returning to ABC. We fixed 300 m of rope on the traverse after negotiating the icefall, and left some stuff at the last point little below the hanging glacier of P. 6041 m.

Now it was time for all of us to move on the 1000 m of rope, which we had been fixing since 30 August. We reached the dump point, but the weather started worsening with white-out covering the slopes of the mountain. Now an avalanche-prone slope and a runnel were ahead of us. We had one 150 m rope some of us would go ahead, anchor the rope, then others would move on it. It took us a long time to repeat the process three times in order to reach the summit camp at 6.00 p.m.

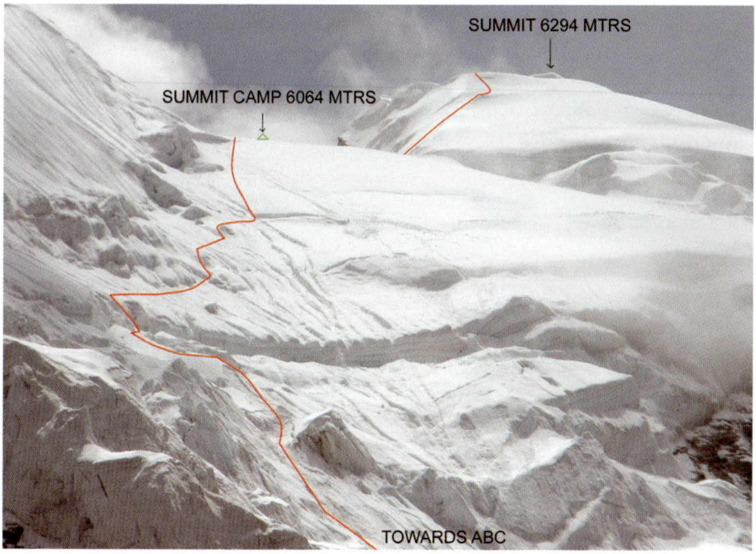

Route through the icefall to the summit. (Dhruv Joshi)

The route till the summit camp

From ABC the first section was a gradual walk on glacial turf towards southwest connected to a 45 degrees and 100 m long avalanche chute followed by a rightward traverse of 20 m connected with a 35 to 40 degree slope of 250 m of long rocky ribs and a 100 m long rightward traverse linked the path up to the hanging seracs.

We then arrived below the huge bergschrund below P. 6041 m. It was connected through ice-snow bulges and filled with avalanche debris coming down through various runnels.

The third section started above the bergschrund. The route consisted of rightward traverse along the left side of bergschrund towards west. The 500 m traverse on the snow slope was prone to avalanche and a complete series of runnels below the P. 6041 m met up a huge crevasse created by the slope coming down from Kuchela Dhura but connected through an avalanche channel to the higher fraction of a similar slope. The traverse was dangerously covered with powder snow.

The fourth section was through more runnels that led the path to the camping site of our summit camp at 6064 m. It was extremely close

Crossing the icefall, Lawan gad seen below. (Wallambok Lyngdoh)

to the col formed by the rutted ridge of P. 6041 m and the slopes of Kuchela Dhura.

Summit camp to summit

Chitramohan, Wallambok and I were a bit exhausted from the last day's tedious climb. So we all decided to rest at the summit camp. We decided to go for the summit early next morning.

We left summit camp at 04.00 a.m. It was foggy as we started moving. We took 100 m of rope with us for belaying and for securing the route. Above the summit camp a 100 m long steady slope adjoined the col between Kuchela Dhura and P. 6041 m in the westward direction. One could see the Shalang valley on left side. The second section was a 150 m long rightward bend and a 50 to 60 degrees slope joined next two consecutive 45 degrees humps of 20 m and 50 m respectively. After reaching the top of the 50 m hump there was snow turf and on the southwest of it was the final summit hump. A moderate slope of 30-35 degrees and 150 m guided us to the summit.

All four of us reached the summit at 08.20 a.m. We stayed on the summit for long waiting for the weather to clear for some good pictures. We

Kuchela Dhura seen from summit camp. (Dhruv Joshi)

plotted the whole route from summit to the base camp with the GPS. We returned at 05.00 p.m.to the summit camp.

Summary

First ascent of Kuchela Dhura (6250 m) in the Lawan glacier of the Kumaun Himalaya achieved by a team from Indian Mountaineering Foundation. The mountain was climbed on 6 September 2013 without the help of any high altitude supporters.

4 Nanda Devi East

Anindya Mukherjee

There are mountains, particularly in some ranges of the Indian Himalaya; that have grown taller in our minds. Many of them are made taller, tougher and almost legendary by an intricate veil of mythology around them. Add a rich history of heroic explorations, epic ascents and a liberal dash of romanticism to that and you have Nanda Devi. And for reasons known, yet unexplained, plutonium or pollution, political or practical; Nanda Devi and its sanctum sanctorum remain out of bounds for less fortunate alpinists like us. The sanctuary is closed since the late 1982 and it seems less than likely that its doors will open in our climbing lifetime! There is however one consolation - one can attempt to approach its mountains from outside, by climbing the outer rim of the sanctuary.

Sometime in the late 2011, I started dreaming of Nanda Devi (7816 m) and a possible attempt to its East summit (7434 m). It all started with an email. I wrote to Roger Payne. Within a few days, I received an encouraging, informative reply from him. It is thus our pilgrimage to the eastern summit of the Blessed Goddess was conceived.

Introduction

> At almost 24,390 feet, Nanda Devi East is bigger, harder and higher than Nanda Kot, and is a sister summit to Nanda Devi, which lies roughly one mile to its west. ..... Tenzing Norgay, of Everest fame, stated it was the hardest thing he'd ever climbed. Our attempt was the most intense mountain experience I've ever had.
>
> Pete Takeda, 'The Secret of Nanda Devi', *Rock and Ice.*

Tom Longstaff in his characteristic boldness had climbed the Nanda Devi Khal (5910 m, more popularly known as Longstaff's Col today) back in 1905 and following his clue the daring Poles achieved the first ascent of Nanda Devi's east summit in 1939.

The most spectacular of climbs on Nanda Devi East happened in 1994. Roger Payne and Julie-Ann Clyma did a truly Longstaff style ascent of the south ridge. There have been a few large and medium sized expeditions after Roger Payne and Julie-Ann's climb. All of them failed. The only significant attempt in the post Roger Payne era, would be Pete Takeda's light weight and fast attempt on the south ridge (2005)[1].

I always felt inspired by Roger Payne and Julie-Ann's climb and their style in particular. Sometime in July 2012, during my Trans-Africa bicycle journey, I got the heartbreaking news of Roger's passing away!

Next morning, I had a clear purpose forming in my mind. I would climb Nanda Devi East in 2013 and I would dedicate this to Roger Payne. His partner and he had set the highest standards of excellence in alpinism on this particular mountain. I am sure Sunanda Devi[2] (as some of the Kumauni's would rather call Nanda Devi East) was not happy either when she heard of that avalanche off Mt Maudit!

Choosing a Style

In March, 2013, less than two months before we were to set out for our mountain, came the mail from Pete Takeda. From the very beginning, we never took Nanda Devi East casually! But, as if being serious was not enough, Pete's words hammered in reality and pretty much so in the right places. I am never in favour of a big team working up a mountain, fixing up ropes, ferrying loads, setting up camps etc. I do not like that mammoth, slow, prehistoric style of 'conquering' a mountain. Apart from our personal preferences, we would never be able to raise the sort of funds needed for such a style. But at the same time, I do not rate myself as a super alpine style athlete as Roger Payne or Pete Takeda! Therefore, I looked for a path in between. We would go expedition style up to the Longstaff's col and adopt Alpine style techniques beyond that.

The path in between and the team

Ours would be a small team, consisting of close friends, working in semi-alpine style, fixing and re-fixing where necessary. We would set up camps, allowing us only two days of ferrying up the 900 m snow face to the Longstaff's Col and then employ alpine style movement swiftly up the south ridge. We would allow ourselves about two weeks, base camp to base camp, to do our job!

---

1. *http://www.seattlepi.com/local/article/Proof-of-plutonium-in-climber-s-sample-1232287.php*
2. http://en.wikipedia.org/wiki/Sunanda_Devi

Nanda Devi East (7434 m). (Anindya Mukherjee)

I have been climbing in the Indian Himalaya for the past decade with Thendup Sherpa. He brought along Karna Bahadur Rai from Darjeeling. And for the load ferries up to the col, we recruited three HAS, namely Kiran Chetri, Lhakpa Sherpa and Temba Sherpa. I wanted at least three more climbers to be included in the team, thus giving us an opportunity to split into two separate summit parties as and when necessary. It was then I invited Ananth H Vishkantiah, Suman Guhaneogy and Aloke Kumar Das to join the gang and the team was complete. With six climbers and three high altitude supporters, we could almost call ourselves the magnificent nine!

On 04 May 2013 we started our trek to the base camp, Bidalgwar. For the next five days we trekked on a nicely built trail, spending nights in Rupsiabagar, Nahar Devi, Martoli, Naspanpatti respectively and reached Bidalgwar (4300 m) on 08 May 2013. Next day was spent in organising loads and checking gear and on 10 May, all nine of us did our first ferry to the foot of Longstaff's Col. At around 4900 m, a snow covered moraine hump would become our advance base camp in a couple of days. But from 11 May to 13 May, for three days and nights it snowed continuously. Helplessly we observed and accepted the fact that we were now left with only 12 days to climb Nanda Devi East and come back down to our base. Not a very comforting thought!

Tackling south ridge. (Anindya Mukherjee)

Looking to Lawan valley from Camp 2. (Anindya Mukherjee)

The col and the giant gendarmes

During the next two days, we ferried up and positioned ourselves at the ABC and then after only two team ferries to the Longstaff's Col, on 17 May, we camped on the col (5910 m) itself. It took us 10 hours from ABC with loaded backpacks to reach there. Snow conditions on the slope and the nonstop elevation was tiring as was the continuous rock fall from the southern flanks of the col. This 'col camp' was our Camp 1. The next morning we were awarded with the much coveted view of the inner sanctuary and the main summit of Nanda Devi.

The next two days were spent in bringing up lengths of static rope from the lower sections of the snow slope below Longstaff's Col and fixing them on pitches of the three giant pinnacles standing as sentinels of the south ridge. We found very little snow and ice on the rock pillars, especially so on the inner sanctuary side. This made our climb interesting, but thankfully without any incident. From top of the first gendarme it is a relatively straightforward yet exposed crossing to the second pinnacle. From the top of the second pinnacle an abseil brought us down on the south ridge and then a short climb to the top of the third pinnacle.

On 18 May, we had left the giant gendarmes behind us and were now very strategically located on the south ridge. It was about time and place to establish another camp. Camp 2 (6200 m) was thus pitched on the very narrow south ridge, metres beyond the third pinnacle, with great views of Nanda's main and east summits.

Over the next three days we, relocated Camp 2 to a buttress after a steep 25 m ascent. We fixed ropes on higher sections of the south ridge and finally on 22 May, after crossing a few rock and snow steps, a gentle snow arête appeared before us. We kept moving up the snow arête and stopped just before it merged into a snow shelf. We dug ourselves a platform and established Camp 3 (6600 m).

Summit day and a failure

On 23 May, eight members started as two roped up parties (Lhakpa Sherpa had stayed back at ABC). At the very outset of our summit attempt that morning, the one striking factor was the wind. It was so strong almost like a gale. We had just this one day at our disposal to give the summit a shot. Two nights of food and fuel was all we had managed to carry on our way up. So we did not really have a choice but to move on! At around 7100 m, just before the quartzite mixed rock sections

Looking towards Nanda Devi Sanctuary from Longstaff's Col. (Anindya Mukherjee)

of the summit pyramid which began at a gradient of approximately 50 degrees, we stopped. Day broke bringing in the slightest of promise of warmth. But by that time, both Aloke's and my feet were frozen solid. Ananth was complaining of a lost pair of mittens (and frozen fingers as a consequence) and Suman was finding his movement too slow. Thendup, as usual, was probably the only person in the party, who seemed unaffected by the environment. I looked around and all the stories of fatalities on Nanda Devi East came rushing back to mind. I had to take a call. Summit or safety? I had a quick discussion with Aloke, the senior most climber and we reached a unanimous decision within minutes. We were turning back. In spite of being a small, lightweight team with limited resources and time; and having overcome all the technical climbing challenges of the formidable south ridge of Nanda Devi East, we decided not to push the last 300 m to the very top.

Over the next two days we retraced our steps and reached base camp on 25 May. We had with us approximately 900 m of static ropes for fixing and we brought it all down.

We dedicate this effort to the great alpinist Roger Payne.

Summary

A lightweight, semi-alpine style attempt on Nanda Devi East by an Indian team in May 2013.

5 Durga Singh Martolia

Chinmoy Chakrabarti

Trekkers to Milam glacier usually do not visit Martoli village on their way since the village is not on the main route and one has not only to take a detour, but also to negotiate a steep ascent. But, I was determined to go to Martoli, simply because, it prides itself on a splendid view of Nanda Devi peaks.

So one day, early in the morning under a clear autumn sky I started from Rilkot village – the last nights' stay, towards Martoli. It was only a trek of four km and the gradient of the slope was not as steep as I had feared. We enjoyed our morning trek on a green grassy slope dotted with alpine flowers in full bloom to give the greenery a nice break, with Gori ganga accompanying us in a gorge which vanished from sight as we gained height. About one and half km before Martoli, I could see the mesmerising Hardeol (temple of god) peak (7150 m). I simply sat down on a flat stone to enjoy the scene.

After a reconnaissance in 1939 and a few serious attempts starting in 1967, the first ascent of Hardeol was made by a team from the Indo-Tibetan Border Police on 31 May 1978, led by S. P. Mulasi, climbing from the ridge connecting the peak to Tirsuli.[1] From this region began the border of the old Johar or Sauka kingdom and went up to the border with Tibet. It was a very prosperous kingdom, mainly, due to a very profitable trade with Tibet. As the kingdom was wealthy, it was repeatedly attacked by the kings of Tibet and Nepal but every time, after a short domination by these kingdoms, Sauka regained its independence. Only when the British conquered Sauka, it lost its' independence forever and after India became free of British rule, it became a very remote part of independent India.[2] As an aftermath of the India-China war, the border with Tibet was closed and so was the business with Tibet. The prosperity of Martoli vanished along with many other high villages of this region as the border suddenly became an un-surpassed barrier. In 1961 Martoli had a population of four hundred and fifty that came down to a mere eighteen in 1981.[3]

---

1. *HJ* Vol 45, p. 71.
2. 'Beyond Milam', *Himavanta*, July 1998; Edited by Kamal Kumar Guha
3. Ibid.

The beautifully carved doors of the ruined houses of Martoli, still somehow standing, tell tales of the old prosperity. As I started to walk, one tall gentleman approached me and asked: "Are you Mr. Chakrabarti?"

Somewhat puzzled, I replied, yes. He introduced himself as M S Bhandari, an inspector of ITBP (Indo-Tibetan Border Police) and in-charge of the ITBP camp there. He explained that he was told to look after us by his commanding officer at Munsiary during our stay there. Well, that's great news! A government servant is certainly lowly paid but enjoys certain advantages!

The gentleman took me straight to the house of Durga Singh, the head man of Martoli village and that's when I first met Durga Singh Martolia; a tall *Rajput* of athletic built with a sharp–intelligent look. He looked maybe 50 yrs of age but the next day I learnt that his age was 65.

Their custom is to add the name of their village to their surname to give complete identity. So Durga Singh became Durga Singh Martolia; meaning Durga Singh from the village Martoli. Since 1982 Durga Singh was the village head-man and he lived there with his elder sister who very lovingly called him 'Duggo'.

Martoli village. (Chinmoy Chakrabarti)

All houses in Martoli were dug-out. Big square holes were dug out of the rocky land and the houses were built inside that hole. Only when a house had two stories, its top floor remained above the ground. Windows were very small to keep out the cold. This hard work to build houses was in order to avoid the very strong wind that blows from Tibet across the pass. The wind starts to blow around 10.00 a.m. and it sounds like a railway engine running at full steam. It stops suddenly around 6.00 p.m.

Durga Singh offered the first floor of his house. The spotless bed was on the floor of his two storied house and was laid down on a soft woolen carpet. 'That was a king's' bed, I thought, because on its head was a small window opening up to a snow peak (Bankatia, I was told) as if framed permanently. The colour of the peak changed as the morning wore into the evening. A garland of dry *Baklo* flowers hung outside the window. This flower has a specialty; it spread its fragrance as it dries up. With the gentle breeze the sweet fragrance wafted in all day long.

My lunches and dinners came from the ITBP camp but I would have breakfast with Durga Singh - a bowl of mashed potato sprinkled with black pepper and very lovingly prepared by his elder sister, who after preparing the food would call her brother 'Duggo the food is getting cold' and we would come running.

There were just five houses and 15 people to administer; so Durga Singh and his sister had a lot of leisure time. Durga Singh had a small plot for cultivation of potato, which took up his early mornings. Durga Singh and I spent hours talking about the Himalaya. He is an expert on Himalayan medicinal herbs and was often invited to Lucknow to advice the UP Government (then UP was not divided) on preservation of Himalayan herbs or to speak at seminars on the Himalayan herbs in different cities of UP. He had two sons and two daughters who were married and well settled in cities. Durga Singh had no financial need to toil on his land but he just could not resist the call of the mountain breeze. So, every summer he would return with his elder sister to cultivate his land and would stay still mid October to bring back his yield to his other house in Munisiary. He single handedly created a forest of Birch trees in the highlands of Martoli village which attracted some musk deer. His only regret was that he was unable to protect these

Houses in Martoli village. (Chinmoy Chakrabarti)

endangered antlers from the poachers. He requested me to speak to the SDM of Munsiary about poaching problem.

On the second day of my stay, very early in the morning, I went to visit the Nanda Devi temple high up on the hill. Nobody was in the vicinity except me and the temple door was not locked; it was simply shut. As I entered the sanctum-sanctorum, an exquisite fragrance surrounded me. It was as if Devi Nanda was sitting on her throne and had hurriedly left on my intrusion; but her sweet fragrance was still lingering. I looked down on the floor and saw that it was covered with dry *Baklo* flowers. I came out of the temple and sat dizzily on its step. Suddenly, a shaft of sunlight fell on the crest of Nanda Devi. It turned golden. Molten gold started to slide down from its crown. I sat there mesmerised.

Nanda Devi must have flown from the temple to that peak and sat down on its crest!

It was now the time to leave of Martoli. Durga Singh's sister gave some mashed potato for the road. Durga Singh walked alongside me till the end of the village. After descending and reaching the main path to Milam, I looked back. Most probably I would never return to Martoli village. Durga Singh also knew that. So I turned back and saw Durga

Singh Martolia, standing on the edge of the ridge. He waved at me; I felt the gentle touch of his rough palm on my shoulder again and a gaze full of love and compassion. Tears filled up in my eyes.

In that instant, I knew, I have a home in deep of the Himalaya, in a magnificent village called Martoli, inside a cozy room owned by my elder brother known as Durga Singh Martolia.

Rato ko chanlewale rahe jaye thak ke jis dam,
Ummid unki meri tuta huya diya ho.
Bijli chamak ke inko kutia meri dekha de,
Jab aasma me harsu badal ghera huya ho

.....Dr. Muhammad Iqbal.

(When the night-traveller is exhausted / his only hope is the light of my broken lampshade. When the stormy-cloud covers the sky / please show him my broken hut by lighting).

Post Script: Durga Singh is still alive and active. He is around 80 now. But every summer he still goes to Martoli to till his land. His forest of birch trees has expanded now to around 100,000 trees with more than a dozen Musk deer. His elder sister is also alive. He lives in Munsiary and remembers every detail of my visit 16 yrs ago. I had a talk with him over his cell phone on 20 April 2013. I told him about my published book on Pithoragarh where he was mentioned in detail and about this article.

He was very pleased.

Summary

A biographical sketch of a resident of Kumaun, Durga Singh Martolia who is an expert on Himalayan medicinal herbs and is also instrumental in growing a large birch forest containing almost a hundred thousand trees around Martoli village of the Milam valley.

6 Exploration of Hanuman Khal

Partha Pratim Mitra

The sun dies down in the west,
The mountains flash with rosy light
And high aloft great Nanda Devi
Glows like a lighted taper.
O Brahma! O Sun born of the lotus
Flower! O glory of the down! We
Yield the thankfulness for this day.

Ven Der Sleen

Nanda Devi region is the mysterious mountain realm of the mighty goddess of Garhwal. Nanda Devi, incarnation of goddess Durga, is feared and revered and loved as no other goddess is.

I have always been fascinated by the mountaineering literature on Nanda Devi and her fairyland. The classic journeys and adventures of the pioneers like W.W. Graham, Dr. T.G. Longstaff, Hugh Ruttledge, Peter Oliver, and David Campbell have ignited my imagination. Inspired by the Garhwali myths related with 'Usha of the Himalaya' and by the epic journey of Eric Shipton my dream finally evolved in an expedition.

A team of five members of Himangan left Joshimath in October 2012 to explore the hitherto unknown valley and a virgin pass beyond north Hanuman basin (at the rim of upper Dharansi plateau). Only Shipton had done major survey work throughout the inner sanctuary. He wanted to summit Dunagiri from Rhamani glacier side but could not. By this valiant attempt he was able to reach a col and observe for the first time an unsurpassable vista. He writes in his survey report –

> ...The following day we managed to reach a col nearly 20,000 ft. high, which connects the southwest ridge of Dunagiri with a peak which on the old one-inch maps bore the strange name 'Niti No.3'. On the northern side of the col the ground fell away with tremendous steepness to the Tolma glen, and we found ourselves looking straight down to Suraithota in the Dhauli valley.[1]

1. 'Survey Work in the Nanda Devi Region' by Eric Shipton, *HJ* Vol. IX, p. 74

Though scantily described from an explorer's point of view, Shipton's exploration fascinated and inspired me. Then I came across H.W. Tilman's article 'Nanda Devi and the sources of Ganges'[2]. Since no expedition had ever ventured into that terrain, our team relied upon Shipton's scanty survey report.

Many mountaineers believed that the upper Dharansi plateau and the north Hanuman basin was hitherto untrodden because these were out of Shipton's survey work and his scanty reports. Harish Kapadia said that he did not have any information on the terrain and encouraged us to consult maps and explore a new area. These words gave us inspiration and an expedition was born.

Our expedition had the following objectives:

1. To get a glimpse of Nanda Devi and her adjacent peaks from the rim of outer sanctuary.
2. To document the hitherto unknown terrain between upper Dharansi plateau and north Hanuman basin.
3. To find out a virgin pass on the northwest rim of north Hanuman basin of Nanda Devi outer sanctuary.

Way to Hanuman Khal, west face of Dunagiri and P. 6029 m in the background. (Partha Pratim Mitra)

2. *HJ* Vol. VII, p. 1

4. To establish a new trekking route between upper Dharansi kharak and Tolma-Tarak valley and survey routes for potential future expeditions to north Hanuman glacier.

On 5 October 2012 we drove up from Joshimath to Lata village. We climbed from the road head to upper Lata village (2317 m) on a well cemented track. We were forced to stay at the primary school of the village due to some porter problems which were resolved soon after.

We visited the famous Nanda Devi temple with Mukesh Singh Power of Tolma village. On the way back a few elderly villagers were surprised to know our plan because they had no idea about north upper plateau of Dharansi kharak. After some argument we assured them that we would go ahead only if we found the terrain accessible.

On 6 October, we started early on Dhauli Ganga's right bank. It was a shaded walk through pine, deodar and after one and half hours we reached at Belta Kharak. After a strenuous climb of 300 m, we finally reached Lata Kharak (3689 m). The trail zigzags up the steep slopes through beautiful forest. On the top of the ridge we came across a forest hut that was built in 1995. From the hut, we observed Ronti (6063 m), Bethartoli (6352 m), Nanda Ghunti (6309 m), Trisul I (7120 m), Dunagiri (7066 m), Ghori Parvat (6709 m), Kuari Pass (4264 m) and other named and unnamed peaks.

On 7 October, leaving Lata Kharak we walked along the grassy juniper filled steep path. Towering above all the peaks to the south stood the white, shining dome of Bethartoli (6352 m). We turned left and in the dense mist, traversed the dangerous cliff and passed through a narrow scree filled slope. Then we negotiated a 600 m sharp rock wall to reach a narrow high point - Sath Kula dhar.

From that point we identified many peaks such as Chaukhamba, Mana (7272 m), Nilkantha (6596 m) and others in the west. Ghori Parvat (6709 m), Hathi Parvat (6727 m), Sapta Parvat of Hemkund Sahib and others were to the north-west. Bethartoli (6352 m), Devisthan I (6678 m), Trisul I (7120 m), Nanda Ghunti (6309 m) were to the south. But the sight that held us spellbound was the massif of Nanda Devi (7816 m) towards the east, queen among mountains, holding her head erect in the vast blue canvas. I remembered the famous lines from *Grongar Hill* by John Dyer:

Now I gain the mountains' brow
What a landscape lies below!
No clouds, no vapours intervene
But the gay, the open scene
Does the face of nature show
In all the hues of heaven's bow.

Then the valley cloud covered the view within twenty minutes. The echoes of roaring torrents of the Rishi below could make most of us falter on the very narrow path. After pulling over 600 m in a grass, boulder filled gully, we reached Dharanshi pass (4237 m), a name given by Eric Shipton to change it from the old name Barf kina Dhar[3]. Across the pass we had to traverse and descend through a broad grassy gully to a grazing ground known as Dharanshi kharak, used in summer by the shepherds of Tolma and Lata. Everyone was exhausted so I decided to call it a day and pitched tents in the camping ground. It drizzled throughout the night.

On 8 October, weather did not improve and thus restricted our views from the meadow. A drizzle started in the afternoon and very soon turned into a thunder shower. In the evening we sat discussed plans to reach Tolma-Tarak valley after finding a virgin pass on the hitherto unknown north Hanuman basin.

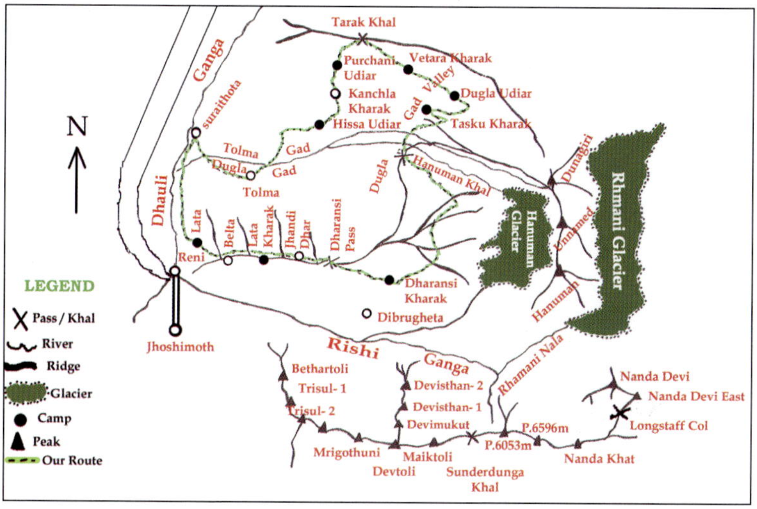

Route Map

3. *Nanda Devi* by Eric Shipton

On 9 October, the vast alps were plastered as it had snowed through the previous night. The weather did not improve. The mountain vistas were covered under a veil of clouds. But we had to get going as every day was crucial and progress, however slow, was mandatory. So we kept moving up north along Upper Dharansi plateau. A steep climb of approx 300 m through a grassy zone led to the top of a moraine wall followed by a left sided delicate traverse for two km. At the end of it we descended through treacherous grass and boulder mixed zone. Next we were climbed and traversed a dangerous cliff.

We then negotiated a few humps in quick succession followed by a steep climb for 150 m. The valley cloud suddenly inched up from southern side. The upper Dharansi plateau is girdled by mountain ridges on three sides. So once clouds enter the solitary plateau, weather remains unchanged until the thicket cloud fully precipitates. Nearly five hours of ceaseless trekking across the ravine of the plateau exhausted us and the dark clouds were terrifying. In poor visibility and blowing wind we kept climbing to a highly elevated ridge. Then we descended to the snow covered boulder zone. We kept traversing to western side on a highly exposed slope of gravel and soft snow that was difficult to hold on to.

Thereafter we started to climb a high moraine wall and an hour later we were upon the high point of the ridge. We were rewarded by a panoramic view on the horizon. Our map suggested that Dunagiri (7066 m) was on our north, unnamed peak (6029 m) and Hanuman (6075 m) on our northeast, Trisul I (7120 m), Mrigthuni (6855 m), Devtoli (6788 m), Maiktoli (6803 m), Devisthan I (6678 m), Devisthan II (6529 m) and many others on the south. We also observed the very distant dense forest territory of Dibrugheta (3350 m).

But most thrilling was the view of the north Hanuman basin with western face of Hanuman (6075 m). This also brought some assurance that we were on the right track.

We observed a horrible blackish cracked part of the Hanuman glacier from which stream emerged was to finally meet the Rishi ganga. I thought of naming the stream as 'Hanuman gad'.

We had thought that if we moved along the stream on our right side towards northeast up to the snout of Hanuman gad, then we would reach the horribly cracked Hanuman glacier. But there was no way to move ahead. This is the terminal part of the north Hanuman basin. The Hanuman glacier encompasses Dunagiri, an unnamed peak and Hanuman which are on a semi circled northern ridge. But the ridge top from where we observed turned to the southwest after adjoining another ridge. Taking bearings from the contour map and compass we decided to avoid the horrible glacier zone and move towards this ridge to proceed. As we moved farther, the dangerous glacier was visible on the right. Then we climbed up a highly exposed steep moraine wall followed by a left side traverse and descended to a narrow boulder zone. Every few second stones and gravel tumbled down.

It began to snow. But we did not waste time and kept moving through boulder mixed moraine zone. As we went higher a depression became quite prominent. This depression looked like a pass linking the southwest ridge and northern semicircle ridge. But it was not marked on the Survey of India map. As this was confusing, we decided to explore the pass.

The natural elevation forced us to climb and traverse the treacherous slope. To our left were two gaps at the depression on the linking ridge and to the right was the ugly black rock and old frozen snow of Hanuman glacier.

The weather could have turned at any time so I sent Mukesh Singh to scout the best route of the two. He reported that the second gap was safe to proceed. So we all followed Mukesh Singh and carefully climbed up a highly exposed moraine wall.

At about 2.40 p.m., we reached the narrow second gap of the pass. It was the moment of delight, because perhaps no human being had set foot on this before us. The wind was blowing and it started to snow. The exquisitely beautiful mountain vista of southern side soon vanished under a veil of strato-cumulous cloud. The unknown destination on the western side was not visible due to mist and clouds. From our maps and compass, we knew this would lead us towards northeast side Hanuman peak and its glacier. I decided to name this virgin pass 'Hanuman Khal' which is at *c*. 5050 m.

Apart from some descriptions by Murray[4] and Chandekar[5] in their respective reports, we have no authentic information about Hanuman peak's adjacent territory. Naturally the north Hanuman basin was mystery to us.

The weather quickly turned into heavy snowfall. In spite of the weather and poor visibility we prepared to cross the pass as soon as possible. The Survey of India map has 'a major blank' between Hanuman glacier and Tolma-Tarak valley. But we took the crucial decision to descend towards the untrodden western territory. After the 'puja' ceremony on the pass, we found a stream emerging from Hanuman glacier and flowing towards west. Another stream which emerged from Dunagiri glacier also flowed west.

We kept descending rapidly towards the uncertainties of the western territory. Bad weather restricted our view and approach. Mukesh Singh tried to find the route by constantly scanning the flow of two streams.

Next we descended approx 300 m and traversed a snow covered spur. Moving an hour we saw two streams meeting and forming a nala. We crossed the terrible current of the nala and reached the base of a slippery rock wall which we climbed. Around 6.00 p.m. we found a small flat area full of shrubs and decided to camp.

This typical shrub is locally called 'Tasku'. I thought of naming the campsite as 'Tasku kharak'. Weather improved giving us partial views of Dunagiri west face and a long ridge which turned towards southwest. According to our map, Tarak khal (4596 m) is on the western end of that ridge.

We started at 8.00 a.m., on the clear, bright morning of 10 October. We descended through a steep rock fall zone to reach the nala. Across the nala we climbed an exposed rock wall followed by a delicate traverse leading to a thick forest. Over the next two hours we had to cross the unrestrained torrent of the nala six times finally reaching the bottom of a steep rock wall which we carefully climbed. The faint trail continued first through rhododendron bushes and then on rocky zone. As the trail turned after a traverse, we reached a rocky slope.

---

4. 'Scottish Kumaon Expedition' by W. H. Murray, *HJ* Vol. XVI, p. 38.
5. 'The Hanuman Expedition' by A. R. Chandekar, *HJ* Vol. XXVII, p. 144.

After negotiating the slope we found a narrow overhang. Meanwhile the weather suddenly turned bad and we decided to camp at the base of the overhang. This type of overhang is locally called 'Dugla'. So I thought to propose the name of the camp as 'Dugla Udiar' and the nala as 'Dugla gad'.

11 October morning was quite clear and we struggled upwards. After we gained about 350 m within a few hours we reached a vantage point from which we rewarded with a magnificent view of a valley. Dugla gad is the main water resource of this pristine hidden valley. I thus decided to call this little valley as 'Dugla gad valley'.

We started our trek at 7.00 a.m. and crossed Dugla gad after the initial descent. We negotiated a a vast boulder zone and then started climbing a steep slope followed by a sharply rising arête of rock. After three hours of continuous climbing we reached the ridge top and were rewarded with a view of the western summit of majestic Dunagiri. We also saw the Tarak khal on the extreme western end of the semi circled ridge which is linked with the western wall of Dunagiri.

The wind was strong so we descended quickly to the other side and reached an alluvial meadow full of shrubs where we decided to camp. This shrub is locally called vetara so we called this campsite 'Vetara kharak'.

On 12 October, we left camp at 9.00 a.m. It took us four hours to climb the three treacherous ridges towards northwest and we reached a small clearing. We immediately started climbing the steep slope of the frontal ridge. At 2.45 p.m. we were finally atop Tarak khal (4996 m). We were getting closer to our goal. We were blessed by majestic view of Dunagiri, unnamed peak, Hanuman, 'Hanuman khal' and 'Dugla gad' valley.

We descended a steep grassy slope. By five in the afternoon the team reached at Purchani Udiar (3500 m).

Morning was cloudy and visibility worse on 13 October. We kept descending towards northwest thorough pine, deodar and Bhoj forests. After four hours of moving from our last night's camp we reached Kanchla kharak (3195 m) for lunch. Then we hiked to Hissa Udiar (2900 m). This was our last camp.

On 14 October we trekked to Tolma village which took two hours and the same day afternoon we reached Joshimath via Lata.

Thus we completed a journey across the untrodden areas between the north Hanuman basin and Tolma-Tarak valley. It was a circuitous route crossing the hitherto unknown Hanuman khal and finally reaching the roadhead of Lata. We hope that our exploration will open up new mountaineering and adventure possibilities in the realm of goddess Nanda Devi.

Summary

Exploration of the north Hanuman basin and first crossing of 'Hanuman khal' in the outer sanctuary area of Nanda Devi by a team from Kolkata in October 2012.

7 Ramblings in Chhota Bangahal

Harish Kapadia

> No scenery, in my opinion, presents such sublime and delightful contrasts. Below lies plains, a picture of rural loveliness and repose; the surface is covered with the richest cultivation irrigated by streams which descend from perennial snows, and interspersed with homesteads buried in the midst of groves and fruit trees. Turning from this scene of peaceful beauty, the stern majestic hills confront us; their sides are furrowed with precipitous water-courses; forests of oak clothe their flanks, and higher up give place to gloomy and funereal pines; above all are wastes of snow, or pyramidal masses of granite too perpendicular for the snow to rest on.
>
> Mr. Barnes, on Dhaula Dhar and Chhota Bangahal: p.9
> *Gazetteer of the Kangra District*, Part I. Kangra (1883-84)

The Dhaula Dhar range, from its junction with the Pir Panjal range, proceeds east-northeast, defining the Ravi and the Beas river watersheds. The range meets the Bara Bangahal ridge and continues northeast till P. 4602 m, (Dainasar peak) at the foot of which lies the historic Dainasar lake. This part of the range is crossed by the old Bajauri pass to a point below the Sari pass to form a boundary between Bangahal and Kullu. From the Dainasar peak, the range takes an almost 90 degree turn to the north to be called 'the Beas Kundi Dhar' and ends near the Taintu pass or Mukerbeh peak. Down below in the plains, the road from Mandi to Kullu-Manali also takes a sharp northerly turn, almost parallel to the range.

This range was recognised by General Cunningham, in his account of *The Great Mountain Chains of the Punjab* as the first part of the chain which designates the outer Himalaya.

'Geography makes history all the world over, and nowhere is this more palpably true than in the Himalaya,' wrote Captain Todd, mentioning legends and history of the Kullu, Mandi and Kangra kings who ruled the states. (*Bruce, p.203*) As the history stretches in centuries it is futile to recall all the kings and rajas. Some salient aspects of the history of these high valleys explain few events.

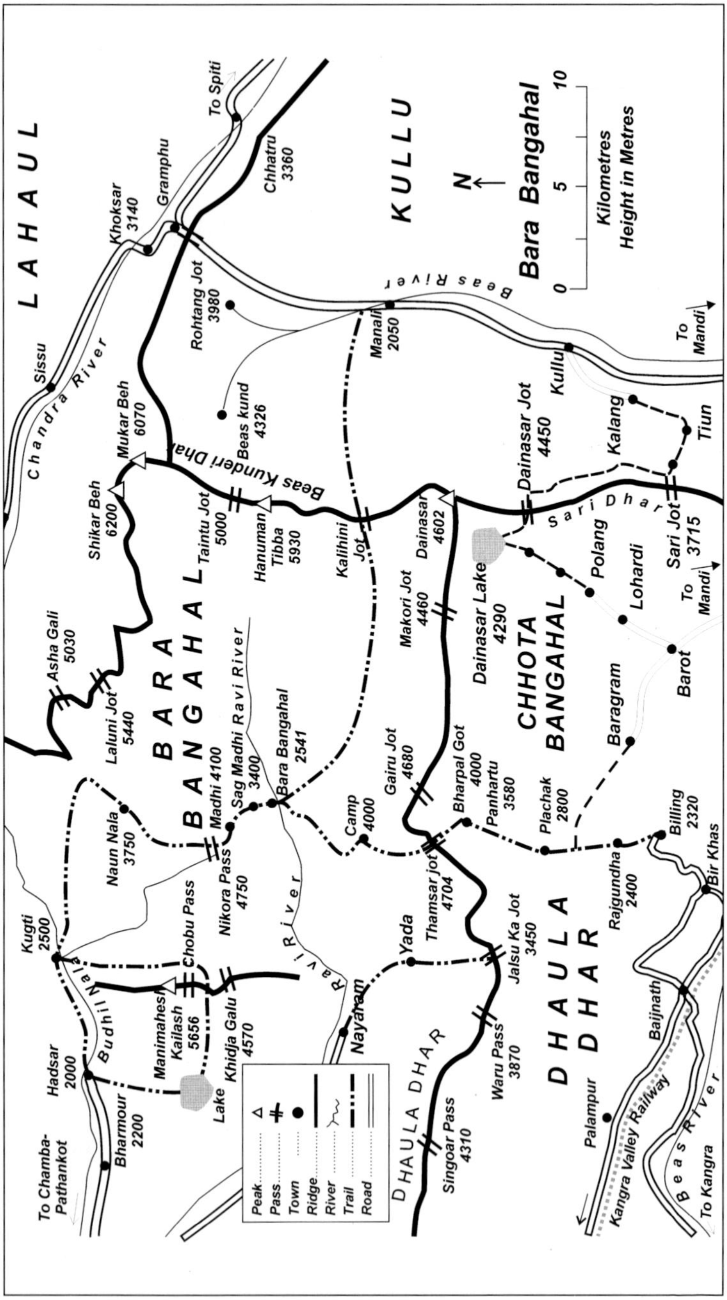

Bara and Chhota Bangahal Map 2013.

One of the many valleys ruled by Kullu rajas was the Bangahali kingdom. Today little is known about how these valleys were given the unusual name of Bangahal. The Gazetteer mentions that 'progenitor of the family *(that ruled Kullu)* is stated to have been one Behangamani, brother of a Kulu prince...... whose family ruled for eighty-seven generations till deposed by the Sikhs in 1840.' 'Behangamani' dynasty was the origin of the title Bangahal. The history of the Rajas and princes is too complicated to be of any use for those interested in its mountains. *(Gazetteer II, p.16)*

Just as several Queens were beheaded in European history, the Bangahali kingdom has a history of many queens forced to perform *sati* (being burnt on the funeral pyre with her dead husband). 'To these instances of cruelty may be added others, fairly numerous according to tradition, where queens were condemned to be buried alive on the faintest suspicion of infidelity'. *(Gazetteer I, p. 28)*

We can safely get away from this gory history with an edict from Mr Louis Dane (former Commissioner of Kullu and one who pioneered the Pin-Parvati pass trail to Spiti). He writes that 'there is a large flat stone below Nagar in rice lands which was used as place for decapitation, and according to all accounts it did not often remain long unused'. *(Gazetteer I, p. 28)*

The Bara Bangahal (larger Bangahal) is an enclosed valley covered by high ridges on three sides, where a passage is possible only through a high pass when not snow bound. The Ravi river originates from many large and small glaciers in the valley and drains west to Chamba, making a deep gorge. Till date except for a scary foot track high above on walls, no other route exists here.

To the south of the Bara Bangahal wall is the Chhota Bangahal, (smaller Bangahal), drained by the river valleys of Uhl and Lambadag. It contains several villages with 'Kothi Sowar' (present day Kothi-Kud) as its principle village. Pirthi Pal, the last raja of Bangahal was treacherously murdered at Mandi by his father-in-law, Sudh Sen, Raja of Mandi. That was the end of a separate kingdom, and Bara, Chhota and Bir Bangahal became parts of the kingdoms of Kullu, Mandi and Kangra. *(Gazetteer II, p. 20)*

Chhota Bangahal contained several valuable metal and ore mines. The early Gazetteers, in 1882-83, mentions eight iron mines yielding 90 'mounds' a year. The mining extended for several kms along the banks of the Uhl river. Some of the ore from these mines was sent to England in 1858 and this variety was found to be as excellent as the product of the best mines of Sweden. In the lower valleys there were salt mines at Drang and Komadh.

The legendary soldier-mountaineer Lt Col. The Hon. Charles G Bruce was one of the first mountaineers to travel in Mandi and Kullu in 1912, with his wife and 'Fuhrer' (oh, no this was far too early for Adolf Hitler, he was accompanied by a Swiss guide named Heinrich Fuhrer!). Being on leave Bruce travelled leisurely through the foothills observing the Dhaula Dhar and Bangahal range. He complained though, 'How often I felt like the retired Colonel who used to engage a bugler to play the *reveille* outside his door every morning at six o'clock, and then cuss him freely and go to sleep again.' *(Bruce, p 2)*

In 2004, I had been to the inner parts of the Bara Bangahal[1] . Scared to cross the Ravi gorge, rightfully, we exited by challenging Nikora pass to Kugti and a long trudge to reach Chamba. Now after nine years I was back with two companions older but fitter than me at 68!

We decided to explore the charms of Chhota Bangahal. At first we entered the Uhl river valley from Baragram to Thamser pass, quite literally crossing thousands of sheep returning from higher pastures. A gaddi coming from the opposite direction will delay your walk as they have the right of passage! It rained every evening but not enough to be threatening. We discussed a plan of traversing from high pasture land of Panhartu to Lohardi over a high route. It proved impossible to get a guide who would lead us over a complicated maze of valleys. Returning to Rajgundha (2450 m), the former seat of the Bangahali kings, we enjoyed few days at the lovely forest rest house and explored a two day walk across the high ridge in the west to hot springs at Tarwani, at the foot of Jalsu pass.

We retreated to Baragram (2370 m) and by bus to Barot. Barot with Multhan, across a bridge, is developing into a tourist resort for vacation hungry people from the plains. Trout fishing here is in plenty. Now we entered the Lambadag river valley. Lohardi (6 km) was a largish village but the road carried ahead due to a Hydro project, till the foot of Polang. As we stood on the road with all our luggage, ladies returning home

[1]'The Bagpipe Treks' by Harish Kapadia in *HJ* Vol 65, p.91.

enquired and went out of their way to help us to get mules organised for our trek. At head of this valley were Gairu Jot (4680 m) and Makori Jot (4460 m) passes crossing into the Bara Bangahal and hence the traffic of returning gaddis continued. On the west near the villages of Chherna and Shatyar is a beautiful forest rest house which many shikaris and trout fishing parties have enjoyed.

The ladies while helping us, narrated the story of the Dainasar lake, where we were headed. The lake is at 4290 m and is on the divide between the Chhota Bangahal and Kullu. Locals make a pilgrimage once a year in September to take a bath which is associated with fertility. A man from Polang reached the lake, leaving his pregnant wife at home. He prayed for twins while taking a bath in the lake and he deposited two golden earrings as offering. His wish was granted and he was blessed with two boys after few months. But a thought occurred to him that his wife was pregnant even before he went to Dainasar so gods had not played a major role! He went to the lake trying to claim gold rings back. He found them easily, blood red in colour, and reached home to find the twins dead for this offence. Since then Dainasar lake is a place for fertility, even if you have not taken any action before going there! However the true importance of the lake and the peak lies in its location at the Dhaula Dhar knot, where it ends and Kullu range start.

On 3 October 2013 we walked leisurely through Polang observing activities and three-day celebrations of Mahatma Gandhi's birth anniversary which falls on 2 October. Soon we were in wilderness and

Dainasar lake (4290 m) in Chhota Bangahal. (Kanu Pomal)

'Brocken Spectre' at Sari Jot. (Harish Kapadia)

proceeded along a nala. After a camp we crossed a bridge and from here it was 1000 m (five hours) steep climb, mercifully on a zig-zag trail. Mules had to be unloaded at two places to make them cross. On top of the ridge was a vast camping ground of Panyali Got (3320 m) but water was some distance away.

The trail continued, climbing but now gently. As we gained altitude, trees disappeared and vast open valleys allowed for a great view. Without stretching ourselves a lot we were at Marhi camp (4000 m) one of the many possible camps before reaching the lake. That evening nature presented a grand show of clouds, fog, distant mountain views and deep valleys.

L to R Indrasan and Deo Tibba. (Harish Kapadia)

6 October was going to be a long day, but it turned out far longer than we had anticipated. At first we climbed to the lake, a lovely bowl at foot of the Dainasar peak. There was a small temple and place to worship. There is another lake higher up

Peaks of Kullu, Dibibokri and Parvati area seen from Sari Jot (3715 m). (Harish Kapadia)

across the ridge but we chose not to visit it. Soon we turned southeast and crossed Dainasar pass (4450 m) to the Sari dhar in the east. This ridge descends south to Sari Jot, forming the boundary between Kullu and Bangahal. As the ridge has a steep drop, the trail proceeds in a roundabout way. Huge rock slabs near the Dainasar pass would keep rock climbers busy for months if they ever come. Soon thick fog enveloped us and the confusion began. Mules had followed another route from the lower camp and we were to meet up at Sari Jot (3715 m).

We followed the guide who now sheepishly told us that he had not been on this trail before but had discussed it with a gaddi well before embarking. Soon we were lost and Vijay, as he is known to, separated from the party and rushed ahead. Exposed traverses, steep rocky descents and shouting for Vijay was the order of the day. Finally we hit the trail but by then Vijay was shouting from almost half way down the valley. Our guide showed lots of energy and finally got us all together, almost at dusk. We decided to wait at a safe place and finally at about 8 pm in pitch darkness, the guide returned with our porters, some torches and water. It was 13 hour day. Tea, hot food and tent never sounded better.

As observed, Sari Jot and Dainasar pass are on the same ridge, which is the divide between Bangahal and Kullu hence a grand view of Kullu peaks opened out the next day. Indrasan, Deo Tibba, Inder Killa, Parvati Parvat, and several other peaks were in view. Sari Jot, on a sharp ridge, also offered a range of 'Brocken Spectre' as the strong sun of Bangahal projected us on the clouds in Kullu. To see ourselves projected in a circular rainbow and move with it is, if you pardon the pun, a moving experience.

On 9 October we descended 1500 m on a far better trail to Kullu, passing gorgeous forest, grazing grounds into a deep valley. Night was spent at Tiun and we continued our descent with children going to school at Kalang, further 1000 m down.

As we reached the roadhead, now the *reveille* was playing in our mind, calling us to civilisation and its temptations.

Members: Harish Kapadia (68), Kanu Pomal (69) and Vijay Kothari (70)

Period: 20 September to 11 October 2013

References:

1. *Gazetteer of the Kangra District, Part I.* Kangra (1883-84). (Reprint Indus Publishing Co., New Delhi, 1994)
2. *Gazetteer of the Kangra District, Part II.* Kulu, Lahul and Spiti (1899). (Reprint Indus Publishing Co., New Delhi, 1994)
3. *Kulu and Lahoul*, by Lt Col. The Hon. C G Bruce. (Edward Arnold, London, 1914)
4. *Into the Untravelled Himalaya*, by Harish Kapadia, p. 149. (Indus Publishing Co., New Delhi, 2005)

Summary

A trek from Chhota Bangahal to Dainasar lake (4290 m).

8 Khhang Shiling and Other Climbs in Bara Shigri Glacier

Cosmin Andron

During 20 July - 10 August 2013 we participated in 'Climbathon 2013', an event organised and sponsored by the Indian Mountaineering Foundation. We were three people from Romania: Mihnea Prundeanu and I - both guides and Vasile Dumitrica, a student. There was one instructor from Scotland/USA – Dr. Paul Schweizer and the rest were from India.

The event took place on the Bara Shigri area and had the objectives of exchange of information between Indian mountaineers and Western mountaineers and focus on alpine style ascents of several peaks in the area. With a total of seven Indian and three foreign instructors and over 60 participants, of which, besides Indians there was one from Brazil, one from Romania and one from Italy, this was quite a large and complex event.

After initial acclimatisation on way to and around the snout of the glacier we moved to set up a base camp on the Concordia via an

Khhang Shiling from Concordia at the Bara Shigri glacier. (M Prundeanu)

Khhang Shiling summit slopes. (Prerna Dangi)

intermediate camp. After three days of preparations and ice climbing training we split into several teams, each with one instructor in charge. Each team was allotted to a group and had chosen an objective.

Our group initially chose Shigri Parvat (6536 m) and we were supposed to attempt it from the southeast face by four separate lines - one team per line. The three teams led by, Prundeanu Schweizer and me were planning to climb alpine style while the fourth team led by Kusang Sherpa was planning to fix ropes on their chosen line. Arriving on location first the team led by Kusang Sherpa deemed the original line they chose too dangerous and proceeded to fix lines across the itinerary envisaged by us. Once our teams arrived at ABC a day later we deemed the climb unsafe in the given conditions. As each team leader was free to choose their objectives and having resolved that the group I was leading would not benefit from joining Kusang Sherpa's already numerous group up the fixed lines I communicated to the Concordia BC that I was reserving my decision until later that evening.

Upon an analysis of the terrain and available options I decided to switch objectives and focus instead on an ascent of Khhang Shiling

1. First ascent of this peak was made by an Indian team led by Divyesh Muni in 2004. For More details refer 'Khhang Shiling – Snow Mountain of Four Ridges', in *HJ* Vol 61, p. 173.

(6360 m)[1] by its northeast face (right opposite the original objective). Prundeanu and his team decided to join me as the line he originally chose was threatened by rock fall. Schweizer decided to wait and maybe attempt Shigri Parvat later as he was still hoping for an alpine ascent of that peak. His team got divided between Prundeanu's and mine with one member joining Kusang Sherpa and one member choosing to remain in the camp. On 4 August we left the ABC around 2.00 a.m. and reached the summit around 7.30 a.m. by two slightly different variations of the same route, with nine people summiting. We were informed that this was a first ascent of the face. We returned to Concordia BC on 5 August. The summiteers were: Cosmin Andron, Mihnea Prundeanu, Vasile Dumitrica (Romania), Angel Robledo (U.K.), Karan Kaushik, Bharat Bhusan, Bhupesh Kumar, Subrata Chakraborty, Prerna Dangi (India).

Schweizer did not have the opportunity to attempt Shigri Parbat and returned with us to BC. During the same period other Indian teams using fixed ropes reached Shigri Parbat, Kullu Pumori and made two first ascents of two other peaks.

On 9 August, following the Climbathon event and after our return to Manali, Mihnea Prundeanu, Vasile Dumitrica and I attempted Kullu

Climbers on the right hand variation of Khhang Shiling. Kullu Pumori in the background. (Angel Robledo)

Kullu Makalu. (Cosmin Andron)

Makalu (6349 m) by its unclimbed north face. Our attempt was thwarted by a massive rockfall that hit us while nearly half way up the face. While I sustained only minor hits on my arm, back and legs and Vasile got away unscathed, Mihnea had a serious spinal injury which had us rushing down the face and crawling through the night to the safety of our BC. Later diagnosis back in Romania showed a hairline fracture of L5 vertebra and a damaged sciatic nerve but with a good recovery prognosis.

Both climbs were made and attempted in alpine-style. Luckily all three of us ended up enjoying beers in Manali several days later.

Summary

Khhang Shiling (6360 m), new route by northeast face and an attempt on Kullu Makalu, also known as Lal Quila (6349 m) by a small team from Romania.

9 The Prow of Shiva

Mick Fowler

Paul Ramsden and I were ready to go. Permits had been granted, tickets bought and bureaucratic hurdles overcome. Pre-expedition excitement was building. And then, the month before departure we received an e-mail from Andrey Muryshev, the leader of one of only two expeditions that had enjoyed a good view of our intended objective, the Northeast Pillar, or Prow, of Shiva (6142 m).

'Frankly I cannot imagine how you will do it. Do you mean northwest buttress? It is *c.* 700 m of climbing after the col and it is northwest side in October – all the rock will be frozen. From the other hand, the ice will be scarce as the buttress is very steep. So it will be very hard dry tooling and very hard protection. I saw your route on Siguniang – it is much easier.'

And Bruno Moretti, the leader of an expedition that had a good view of the east side in 2010 had already suggested that the rock was likely to be terrible.

All in all it didn't paint that positive a picture. But the photos Andrey and Bruno had kindly shared with us spoke for themselves. Shiva is an isolated 6000 m peak in the Pangi district of the Indian Himalaya. Bruno felt the Prow stood out as the best line in the area and Andrey's e-mail ended by saying he thought it was 'inspiring'. That was good enough for us. I did wonder if we were up to it but loose rock, difficult climbing or whatever the line was such that we had to give it a go.

The Pangi district borders Kashmir and the troubles there have impeded access until recently. Now though the political situation is better, roads are being blasted ever further into the Indian Himalaya and mountains such as Shiva are accessible in 30 day trips from the UK. And to further ease matters in 2012 some economy flights came with a free 46 kg baggage allowance. Times have certainly changed since my early trips to India where the first couple of days would be spent enduring bureaucratic misery retrieving freighted equipment from the Delhi customs warehouse. With a 46 kg allowance all our equipment fitted easily in the hold and with mountain gas cylinders now being available in India life is certainly a lot easier than it used to be.

Prow of Shiva. Route of ascent (Mick Fowler)

Steve Burns and Ian Cartwright made up our four man team and in late September we were marvelling at the new competition standard climbing wall at the Indian Mountaineering Foundation building before heading off on a night drive to the 'honeymoon town' (as our liaison officer called it) of Manali in the Himalayan foothills. From here it was a long day's drive over the Rohtang pass and then along a remarkable road down the Chenab valley to eventually turn off and reach the roadhead at the small village of Saichu. As far as we knew we were only the second mountaineering expedition, after Bruno Moretti in 2010, to visit this valley.

The weather seemed set fine and as water levels were low, we had understood that it would be possible to use mules or horses to carry our equipment all the way up to a base camp at 3900 m below the east face of Shiva. As ever though planning on Himalayan trips has to be able to cope with a little flexibility. The first day was idyllic trekking through grazing pastures and deciduous forests but by evening the valley forked and it had become clear that the fork we needed to follow, the Tarundi valley, was so full of dense, shrub like bushes that the horses could go no further. Porters were summoned and by a combination of portering and us load carrying base camp was established in a fine spot just above the bush line. Above us Shiva was clearly visible with the Prow looking even more inspiring than we had expected. We had clearly come across something rather special. And, through the binoculars, it did seem as if the rock on the steep section might not be as bad as we had been led to expect.

First though we had to acclimatise. And an unclimbed 5500 m summit on the ridge leading north from the Prow gave an obvious objective at just the right height. The views from the top were mouth watering. We could see that there were a few uncertainties about getting onto the crest and up to the foot of the steepest section but the most interesting discovery was that the Prow itself really did appear to be composed of good quality granite with occasional good cracks. It was one of those objectives that we both knew we would have to come back to if we didn't get up this time.

Back at base camp Brittam our cook and Devraj the kitchen boy were probably the finest we have ever had on a trip and after a rest day during which we were fortified with large quantities of fine food it was time to get used to a diet of small quantities of mountain food and live

Paul Ramsden acclimatising on P. 5500 m. (Mick Fowler)

Mick Fowler with head-on slope approaching crest on Day 3. (Paul Ramsden)

Crossing the icefall. (Mick Fowler)

PR travesing just below steepenng of buttress

off blubber accumulated over the year since my previous Himalayan exertion (two years in Paul's case).

So on 5 October, after tackling the complex glacial approach we pitched our little tent at the foot of the east face. The obvious first difficulty was getting over a bergschrund to establish ourselves in couloirs leading to the crest at the point where it steepened towards the Prow. Juggling with job, family et al means that I only manage one mountaineering expedition per year and it always takes a pitch or two to settle back into the swing of things. Being an east face an early start was necessary and 1.00 a.m. found me experiencing a harsh re-introduction to mountaineering by tackling a vertical shale wall with a slanting, overhanging wall of iron hard ice on top. I certainly felt fully exercised by the time I pulled out into an ice runnel on the slope above. Back in England I had envisaged that it would be straightforward to gain the crest of the buttress below the Prow. In reality though the angle of the approach face was considerably steeper than I had expected and the combination of awful rock and powder snow made for sections of tricky, poorly protected climbing. The amazing variation in the snow depending on the direction of the slope added further to the challenge with north facing areas being covered in remarkably deep and remarkably steep powder.

'It's rubbish here. Just a collapsing knife edge'

Paul was in the lead and his comments were not encouraging.

We had reached the ridge at 9.00 a.m. just as the snow was beginning to soften badly.

An interval followed during which I could pick out some energetic 'a cheval' activity going on as Paul moved slowly to a slightly more amenable spot. By the time he had brought me up the Ramsden face was smiling.

'It's soft right the way down. Could be good for the tent.'

And it was. By about 10.00 a.m. a section of the knife edge had been flattened, the tent pitched and a belay of sorts constructed around a large snow bollard. It looked as if the rope would probably cut right through it like a cheese wire if put under any strain but that was not the sort of thought that we wanted to clutter our minds with as we lay relaxing and reading our books for the day. Above us the soft snow ridge looked challenging but the mixed buttress which started perhaps 150 m above us looked to offer perfect, safe climbing in a position of the kind Paul and I scour the earth to find.

The leaning tent bivi. (Mick Fowler)

By mid afternoon the following day the soft snow was behind us, we were underway on the Prow proper and had reached a small balcony just on the east side of the crest. Below it the ground overhung steadily for at least 500 m while round on the cold northwest side the way was barred by a completely smooth 75 degree rock slab with an intermittent thin covering of verglas and powder. The only way up seemed to be an overhanging fault line above the balcony, but that could be seen to cross the crest after perhaps 10 m and then disappear under the verglas and snow sticking to the slab. Paul and I prefer to carry our sacks wherever possible but this was clearly one of those possible 'stopper' pitches where a sack-less leader was called for. I watched nervously as Paul aided the overhang and moved more precariously on up the fault-line. Looking around there really wasn't anywhere else obvious to go. Bolts could solve the problem but we both feel strongly that they have no place in exploratory mountaineering.

We were right on the crest of the buttress and if we couldn't do this pitch then that could well be the end of our attempt. All our dreams would end here. My heart was in my mouth as Paul moved up round the crest but he just kept on going. Brilliant! The fault-line clearly continued in some way up the slab and eventually the shout came for me to follow. The fault did continue but in a distressingly thin manner. It was a fine lead by Paul and it left us with a good feeling. We had agreed to explore this section in the afternoon and return to bivouac on the balcony ledge but it still felt somehow strange to be leaving the sacks and following the pitch without being weighted down by a sack. We tend to climb strictly with alternate leads and with enough daylight remaining it seemed sensible for me to lead another pitch even though this would leave us with two pitches to jumar up in the morning. And we both hate jumaring. It feels unethical somehow; all that hanging about on ropes instead of climbing.

Weather wise it seemed we had timed things just right. Every morning the sky was clear and the weather glorious and every afternoon it clouded over with a gradually increasing amount of snow. It was very cold but the morning sun was great for thawing us out and infusing a sense of great happiness. The day above our nose to tail balcony bivouac was one of the finest I have enjoyed in the mountains. Out there on the crest of the most eye catching feature in the area, the climbing started with absolutely perfect ice choked granite cracks and continued with memorably thin ice on steep slabs which were not unlike a lean day on the harder slab routes on Ben Nevis. By late afternoon a nose to tail bivouac ledge again provided a lie down bivouac and the possibility of sound sleep with a perfect view from our bedroom. People sometimes ask me why Paul and I never go for the headline grabbing objectives on bigger mountains. Here, on this wonderful climb in a rarely visited part of the Himalaya we had our answer. Shiva was giving us everything that we could possibly want from our mountaineering – an unclimbed magic line, visible from afar, going straight to the summit of the mountain, an interesting new area for us, no-one else around, a great approach and a traverse of the mountain with an unclimbed descent route in prospect. And all possible whilst still leaving enough time out from our full time jobs of health and safety adviser (Paul) and taxman (me) for family holidays.

The next day was our sixth out from base camp. A few inches of snow fell in the night and it was gradually dawning on me that a descent down our line of approach up the east face would be unpleasant to the extent that, whatever lay ahead, it was better to persevere than descend. More thinly iced slabs followed by wonderful ice grooves and intimidating steep mixed pitches finally led to a snow band beneath the last overhanging wall which guarded access to the summit. From what we had seen through Paul's binoculars the band led rightwards round the crest to steep snow and a way of avoiding the final wall. Snow was falling as I headed off rightwards round onto the northwest side. But

Shiva from the east. (Andrey Muryshev)

Descending southeast flank on Day 8. (Mick Fowler)

luck was against us. A gap in the ramp presented a challenge that would obviously require aid and be time consuming. I retreated forlornly to an increasingly cold Paul. There was no way we were going to get up that evening and we had to admit that an ice choked chimney splitting the headwall was looking a more promising option for the morning.

Squeezed under the final wall was the one ledge on the entire Prow where we could half pitch the tent. Much of the floor overhung space but getting the poles in makes such a difference. Shielded from the heavy overnight snowfall we enjoyed a comfortable night and were ready to tackle the chimney first thing in the morning on day seven. It gave a fitting finale and from a cozy niche belay at the top I soaked up the weak rays of the morning sun and the glorious view as Paul led off out of sight up towards a short ridge leading to the cornice. It fell to me to break through and enjoy the curious sensation of being able to walk around on a flat area. Paul came up and a summit hug was in order.

A descent of the unclimbed southeast flank ended nine wonderful days in the mountains. It had been a fantastic trip and we had completed a climb that we already knew would soak us in retrospective pleasure for many years to come.

Summary

First ascent of the northwest pillar of Shiva (Prow) (6142 m) achieved by a two-man British team in Alpine-style, in a nine day round trip. The summit was reached on 11 October 2012.

10 The Himalayan Bimble

First Ascent of Cha Ri (6046 m)

Douglas Briton

The aim of the expedition was to explore that part of the Himalaya that had received no recorded visits from mountaineers, and if possible to climb an unclimbed peak, P. 6030 m. As this was the first exploratory expedition I had organised to the Himalaya, I sought an objective that, while still exploratory, would minimise our logistical complexity, have the easiest access to medical help should we have a problem, and give the highest chance of success. P.6030 m met these criteria. First of all, it was already listed as an open peak, thus removing any complications around getting permission to climb. The walk from the road to the foot of the mountain appeared to be relatively short, perhaps 12 to 14 km. It started from the village of Gya which has a clinic and a satellite-based STD phone.

Gya itself is just over two hour's drive from Leh with its hospital and airport. These factors all made for simple logistics and for relatively easy access to medical help.

Finally, examination of the area using Google Earth suggested the possibility of routes to the mountain and on the mountain that avoided glaciers and which were not extremely steep, thus maximising our chances of success.

In addition I had visited the area in 2011, and had spent a week in Gya High School, just a few km up the road from the start point in Gya village, and so the team had better and more recent knowledge of this area than of any other part of the Himalaya.

Route to advanced base camp

We set up a base camp at Salsal campsite. This is on the main Leh to Manali road, approximately two km north (i.e. downhill towards Leh) of Gya. We camped here for two nights to aid acclimatisation. We used this time to locate the clinic and the satellite-based STD phone within the village.

We followed the Yabat river all the way to advanced base camp. From Gya to Camp 3 the easiest ground was on the south side of the river. Beyond Camp 3, the river disappears under a boulder field. There is a path through this boulder field marked by small cairns, presumably set in place and maintained by yak herders.

Camps 1 and 2 were south of the river. Camp 3 and advanced base camp were north of the river, in or beside livestock pens that had been constructed from dry stone walls.

Exploring routes to the summit

There were four routes to the summit accessible from ABC: the northeast face, the north ridge, the southeast ridge and the south face.

From ABC, the northeast face and the north ridge looked unstable, with a high risk of pinnacles and other loose rocks breaking from the ridge and falling down the face. We therefore rejected these two route options. We decided to climb the moraine at the foot of the glacier that came between P. 6030 m and P. 6120 m to the south and examine the other two routes more closely.

Finding a safe route up the moraine was difficult. From ABC we selected two routes to try: the northern and the southern fringe of the moraine, where the glacier debouched from between the south face of P. 6030

Cha Ri (right most), Kang Yabat II left of the saddle and Kang Yabat III to the right. (Caroline McCann)

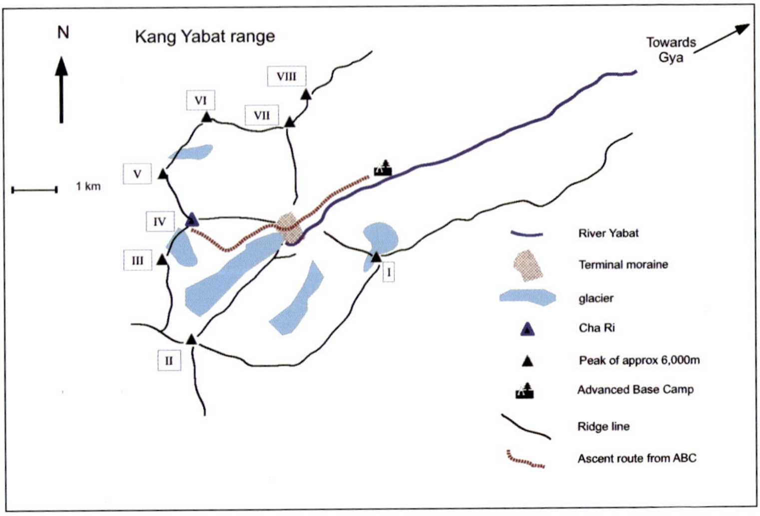

Sketch map. (Douglas Briton)

m and the north face of P. 6120 m. The northern fringe was unstable and topped by some steep piles of loose rock which caused us some concern. Our descent route down the southern edge of the moraine was even worse. This route followed the stream which escaped from the top of the glacier. There was near constant rock fall from the north bank of this stream, though we managed to find a route away from the fall line of these rocks. Nevertheless the descent was fraught, over steeply piled and loose boulders. However this reconnaissance trip was successful in that we saw that the southeast ridge was also unstable, but that the south face had a subsidiary ridge that looked like it might provide a safe route to the top. More, the summit itself was snow-free, and there appeared to be a snow-free route to the top. This opened up the possibility of a summit attempt without the weight of ice axes and crampons.

Another benefit of our reconnaissance was that we spotted what looked like a safer and easier way up the moraine, following a slight ramp to a slight notch on the moraine's top.

The next day we rested and made plans It was clear that at this time Matt and Caroline were stronger than myself, and the best chance of a team summit was for Matt and Caroline to climb as a pair. I therefore decided not to attempt the summit.

Possible routes to summit

The south face: This looked to be a good option, either to the left of the hanging snow field and across the ridge, or using winter gear to climb the snowfield itself. At close quarters though, the snow was wet, loose and constantly sloughing. Climbing it or crossing it would have been difficult and very dangerous.

The southeast ridge: This was our preferred route during planning. On closer inspection it was found to be made up solely of loose rock. There was recent evidence of rock falls so this route was also discounted on safety grounds.

The north ridge: Long and steep, very technical and exposed. This route was discounted on first sight, during the recce. It would be fantastic for future exploration but will need plenty of time and full technical climbing gear.

Summit day

To maximise Caroline's and Matt's chance of success, Wangckuk, Dorjay, our liaison officer Munesh and I carried Matt and Caroline's bags to the top of the moraine, along with spare water and food for them to cache for their descent. Caroline and Matt also carried a small stove so they could melt snow on the mountain.

Getting from the moraine to the south face proved more difficult than thought: Caroline and Matt found themselves on the wrong side of a previously unseen glacial lake. It was severely undercut, eroding constantly and required quite a detour. This delayed them by over an hour. On viewing the face from close quarters the climbers saw that the ridge might be treated as a scramble. They stashed their rock climbing equipment (rope, harnesses, protection) above the glacier and took only their helmets, plus the clothes and safety equipment appropriate for such an ascent.

The lower flanks of the mountain were quite loose scree and rock, but the upper reaches were solid red granite which Caroline and Matt enjoyed climbing. The summit pyramid however, was very steep and loose. The final eight m were too steep and unstable to bear the weight of the team, so the climbers stopped at the highest point they deemed safe.

Kang Yabat I from the north. (Douglas Briton)

The descent was helped by a long scree run from the summit. Meanwhile those of us in ABC were concerned for their safety as storms swept the valley, including a hail storm powerful enough to break up the soil. Fortunately the mountain remained in a small weather window: Matt and Caroline watched storms hit other peaks and valleys but the bad weather avoided them.

The ascent took 13 hours in all.

The next day, the climbers' decision not to ascend the final pile of rocks was validated when Munesh, the Liaison Officer saw a significant rockfall from the summit cone.

The whole range of Kang Yabat and its eight peaks fan the head of the valley.

The glaciers, all feed the Yabat River which flows down to Leh and eventually joins the Indus.

Note on the nomenclature

Our Ladakhi contacts explained that in Ladakh it is the glaciers that are more important than the mountains, because it is the glaciers that supply the water that enables crops to be grown in the summer. What to western eyes may appear to be a range of distinct peaks and glaciers is often given a collective name because it all feeds the one river. Examples include Mentok Kangri I and II, Saser Kangri I, II East, II West, III and IV etc.

We chose to follow this example and name the range of peaks that feed the Yabat river the Kang Yabat range. Our understanding is that 'kang' is a Ladakhi word meaning ice or icefield or glacier.

Discussions with another Ladakhi speaker suggests that the Yabat river gets its name from 'yababs' which is Ladakhi for flowing from ice or flowing from a field of ice or flowing from a glacier, which would match our observation.

We decided to name the actual peak Cha Ri.

'Ri' means mountain in Ladakhi. 'Cha' can mean a large flying bird, and each day we saw a large bird, probably a Lammergeier, fly round the peak. But 'cha' also means tea and we drank a large amount of tea in the course of our expedition.

Note on glacial lake

We discovered a lake on the glacier running along the south flank of Cha Ri. There was very active erosion from the sides of this lake with the sidewalls being significantly undercut.

Given that the surface runoff from the glacier formed a short tunnel before pouring down the corner between the moraine and the north ridge of P.6120 m, there is a possibility that water from this lake could form a tunnel to suddenly breach the moraine, causing a sudden pulse of water to flow down the Yabat river.

In Gya village, many houses have been built very close to the Yabat river. A sudden release of water from this glacier lake may therefore pose a threat to life and property.

Members: Douglas Briton, Caroline McCann and Matt Jones

Summary

A small British team completed the first ascent of P. 6046 m (proposed name - Cha Ri) near Gya in Ladakh Himalaya in August-September 2013.

11 First Ascent of Nga Tsoey Kangri

Kyoto University Alpine Club Zanskar Expedition

Hiroaki Ogihara

Planning of our Zanskar expedition

The Kyoto University Alpine Club was interested in the southern Zanskar region of the Indian Himalaya, ever since we read the exploration article of Reru valley written by Kimikazu Sakamoto who was a senior member of the Academic Alpine Club of Kyoto. We requested a report of his exploration to Lenak and Giabul nala of Zanskar in which his team was planning to find the hidden virgin peaks of the region in the summer of 2011.

We decided to climb the virgin P. 6070 m (L15)[1] in the left branch of Lenak nala in the summer of 2012. Our second target was P. 6080 m (L13) in the same area. Our team consisted of four members, Hiroaki Ogihara (Senior) leader, Yusuke Morimoto (Senior) deputy leader, Yuki Sawada (Sophomore) and Karin Kosaka (Sophomore). Tsewang Yangphel, worked as our guide and he arranged five staff members. Our expedition team had eleven persons including a liaison officer.

Access to base camp in Lenak nala

We flew from Delhi to Leh, with a population of about 50,000 at an altitude of 3500 m. From Leh, we travelled by car for three days to reach Padum, which was the centre of Zanskar. We drove to the roadhead, Dorzong.

03 September: Dorzong to Tsetan

04 September: Tsetan to Testa

05 September: Testa to Shanka. Shanka is the small village with only three houses and is the junction of two valleys, Lenak nala and Giabul nala.

06 September: Rest at Shanka.

07 September: Shanka to camping ground in Lenak nala. We could see the attractive P. 6070 m (L15).

1. *HJ* Vol. 67 p. 246, *HJ* Vol. 68 p. 30

BC and P. 6070 m. (L15)

08 September: Rest at camping ground in Lenak nala. We met a JAC team descending after their first ascent of P. 6165 m (L10) on the right branch of Lenak nala and exchanged information

Climbing activity

09 September: Though it seemed to rain at any time, we departed and set up our BC at around 4800 m on the left branch of Lenak nala. Ogihara and Morimoto went to reconnaissance the route and reached about 5050 m at the foot of P. 6070 m (L15) climbing a steep slope.

10 September: Rest at Lenak nala BC. P. 6070 m (L15) seemed rather difficult and dangerous with steep scree and snow avalanches. We decided to attempt our second target P. 6080 m (L13) instead of P. 6070 m (L15). We obtained permission from our liaison officer.

11 September: Four members, the liaison officer, Tsewang and four staff members moved up from BC to establish our climbing route on P. 6080 m (L13). Soon after leaving BC, we climbed up the moraine of Lenak nala glacier. From 5000 m, we entered the north valley to P. 6080 m (L13) and set up C1 on the right side of the glacier. After taking a break, Ogihara and Morimoto scouted for C2.

By climbing to the tip of the glacier, we got a full view of this peak and the connecting glacier. The glacier was not covered with snow, and there was no danger at all of hidden crevasses. After about 40 minutes climbing, we reached the expected camp site for C2.

We climbed up to about 5500 m and deposited our climbing equipment in the shade of some rocks. P. 6080 m (L13) had less snow than we had expected.

12 September: Rest at BC. The route was clear since there was little snow accumulated on P. 6080 m (L13) glacier. All of us were in good health so we decided to set up C2 next day and then prepare right away to attempt the peak.

13 September: Four members departed from BC. At C2, we placed some slate stones on the glacier surface and pitched our tent over the stones. Sawada suffered from headache and dizziness. So, we decided to go down to C1. It started to get foggy when we reached C1 and it snowed again. In the middle of the night, Kosaka vomited several times. It was clear that she was suffering from high altitude sickness.

Climbing route of P. 6080 m.

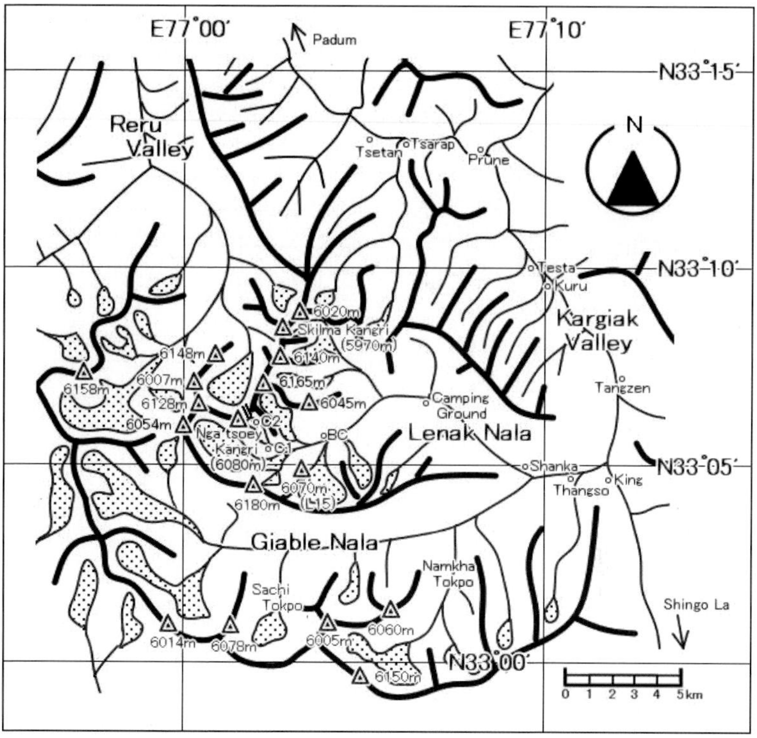

Revised outline of Lenak nala ver 2.0

14 September: We decided to let Sawada and Kosaka descend to BC, though they seemed to be getting better and were disappointed to go down. The weather cleared up soon. Ogihara and Morimoto started to climb to the summit and made an easy ascent to C2 and prepared to ascend the peak under a clear blue sky.

15 September: We overslept a little. Even though the wind was strong in the evening and there were many clouds during the night, Ogihara and Morimoto decided to go for the summit. Morimoto led on the scree slope, avoiding crevasses. It was silent, as the water stream coming from the melted snow was frozen. As the scree slope was rather gentle, we were able to go up quickly. At around 5.30 a.m., the sun began to rise. At 6.00 a.m., we reached the tip of the snow ridge. While the snow was hard, the slope was only about 30 degrees. We zigzagged up the slope and reached the col at 6.45 a.m. Viewed from the col, the ridge to

Looking at the west peak of P. 6080 m.

the east peak was wide open. We walked up along the ridge for about 15 minutes and then Morimoto was the first to stand on the virgin peak 6080 m (L13). The slate scree slope was exposed on the East Peak itself. We relaxed, making a cairn and taking photos.

From the east peak, we could see the ridge from the col to the west peak. It looked more difficult than expected. As we could see two crevasses running parallel to the ridge, we proceeded using our rope on three pitches. As the snow was hard, we had difficulty in setting up an anchor. The west peak of P. 6080 m (L13) was a bare rock peak, where we found something like a cairn. There was no other artificial mark. When we discovered the cairn, we were surprised rather than disappointed.

We went down to the col without using a rope and returned to C2 without crampons. After returning to C2, we decided to stay overnight instead of going down to C1 on the same day, since we were tired.

16 September: We dismantled C2 and headed down to C1. It started snowing after we reached BC. Truly, we were lucky to be able to reach the summit the day before.

17 September: Rest at BC. We decided to try our original target, P. 6070 m (L15) after resting at BC. Though it snowed all day, there was not much accumulation of snow. In the afternoon, we saw some breaks in the clouds.

18 September: Tsewang woke us at 4.00 a.m. We saw everything was covered with snow.

He told us that it had started to snow in the middle of the night and the much snow had accumulated. It was our big regret that we had to give up our original target P. 6070 m (L15). But we made our decision to abandon our plan, because the danger of avalanches and the risk of hidden crevasses were too high. Also, Tsewang warned that we might be confined as our horses would not be able to walk down in the heavy snow if it kept snowing.

Fortunately we could go down safely to Shanka village, where we stayed two nights.

Epilogue

When we visited IMF in Delhi, we paid the additional climbing fee for P. 6080 m (L13) and reported that we saw the cairn on the top of P. 6080 m (L13). We were informed that our's would be recognised as a first ascent if we submitted our official expedition report to them after returning to Japan, because there was no previous climbing record of this peak.[1] We named P. 6080 m (L13) as 'Our Mountain', which translates into 'Nga Tsoey Kangri' in the Ladakhi language.

After returning to Japan, we received an e-mail from IMF stating that 'Nga Tsoey Kangri' has been accepted as the official name for P. 6080 m (L13).

Summary

First ascent of P. 6080 m in the Lenak nala of southern Zanskar on 15 September 2012. The Kyoto University team named the mountain as 'Nga Tsoey Kangri' meaning 'our mountain'.

1. As a cane was found on the summit this climb cannot be called a 'first ascent' even though the first climb was 'unofficial'.

12 Exploration of Zanskar's Pensilungpa Valley

Derek R. Buckle

Introduction

Inspired by Harish Kapadia's article *Zanskar Anyone?*[1], we decided to look more closely at possible climbing objectives that could be achieved from the relatively accessible Pensi la. It was immediately clear in contrast to the dominant Durung Drung glacier, the neighbouring Pensilungpa was relatively unexplored; particularly with respect to the subsidiary glaciers descending from the Pensilungpa - Durung Drung divide. Armed with photographs kindly provided by Harish we chose to look more closely at those unclimbed peaks that could be approached via these side glaciers. Five such glaciers descend from the divide, but as a result of marked glacial recession none now flow continuously into the Pensilungpa but are separated from it by substantial terminal moraine debris. Indeed, remote sensing suggests that as much as 38% of the Pensilungpa glacier area has disappeared over the period from 1962-2007[2] and that the glacial snout has receded some 120 m over the same period[3]. Evidence of previous surveys to map the position of the main glacier snout in 2007, 2008 and 2009 were noted and the current position of the snout was recorded as N33°50.873′ E076°18.907′ at 4640 m, indicating an average annual retreat over the period 2008-13 of approximately 11 m. As the source of the Suru river, the Pensilungpa glacier is an important tributary of the Indus, which it joins to the north of the town of Kargil.

Only Z8, which dominates the north eastern entrance to the Pensilungpa, is well documented. It was first climbed by the Italian husband and wife team of Gino and Silvia Buscaini via the west-northwest ridge in July 1978 and has been accorded a height of 6050 m.[4] Our own estimates, based on sightings from neighbouring peaks, suggest that the true height may be rather lower. We have found no accounts of attempts on other peaks to the southeast of the Pensilungpa glacier.

1. *HJ* Vol. 65, p. 151.
2. A. C. Pandey, S. Ghosh, M.S. Nathawat & R. K. Tiwari, *J. Indian Soc. Remote Sens.*,40, 245-255 (2012).
3. S. Schmidt, M. Nüsser, M. S. Nathawat, S. Ghosh & A. C. Pandey, *Geophysical Research Abstracts, 12*, EGU 2010-13451 (2010).
4. G. Buscaini. *American Alpine Journal,* Vol. 52, p. 614 (1978).

North face of Z8 from glacier 1. (Derek Buckle)

After flying from Delhi to Leh, a two-day drive took us via Kargil to Ringdom and thence to the roadhead a little north of the Pensi la. From here a three kilometre trek took us to a base camp at 4630 m, just half a kilometre short of the current glacial snout. From this camp it was possible to access the glacier proper, and its tributary glaciers to the southeast, by way of the complex true right lateral moraine.

Exploration of the Pensilungpa and its southeastern glaciers

Once established at BC our priority was to explore the topography of the valley; especially that bordering the third tributary glacier from which

Pensilungpa glacier. (Derek Buckle)

we hoped to access our first climbing objective. Two of us therefore ascended the well-defined medial moraine starting immediately above the glacier snout and ascended this until its coalescence with the relatively rubble free glacier at approximately 4900 m. From here we sighted possible objectives from the second and fourth tributary glaciers, but our primary target at the head of the third glacier obstinately remained hidden by a prominent pyramidal peak (tentatively named 'Pyramid Peak'). Indeed, it was not until we subsequently climbed high on this glacier system that we sighted the peak and it was rarely visible from all other vantage points that we visited. For this reason we eventually proposed to call it 'Hidden Peak'. Possibly more importantly, from the lateral moraine we were able to determine likely positions for higher camps and their potential access routes. Meanwhile the second pair followed the true right lateral moraine, traversing beneath tributary glaciers two and three to establish a viable route of access in addition to the identification of possible future camp sites.

As a result of this reconnaissance, and with the assistance of four willing Sherpas, we were able to establish Camp 1 at 4845 m on 11 September and Camp 2 at 5223 m the following day. In continuing brilliant weather Mike, Tony and I set out early on the morning of 13 September to attempt Hidden Peak while Chris, who was not fully acclimatised, remained at Camp 2. Climbing unroped until the gradient increased significantly (to 30°) we were then forced to zig-zag through a maze of crevasses, firstly on the true right and then more

Pyramid Peak. (Derek Buckle)

towards the centre until the slope began to ease off. At this point the north face of Hidden Peak came into view just as the sun rose high enough to greet us. All earlier doubts about whether we had approached via the wrong glacier were immediately dispelled as we headed for an obvious col between Pyramid and Hidden Peaks. From this rocky col a short, steep snow climb up the north-northwest ridge led to a narrow snow arête, which in turn led to a prominent rock outcrop some 15 m higher. Easy climbing gave way to a compact 5802 m summit which we reached around three and half hours after leaving Camp 2 to make what we believe to be the first ascent (Alpine PD+). This vantage point afforded excellent views of the broken south ridge of Pyramid Peak and

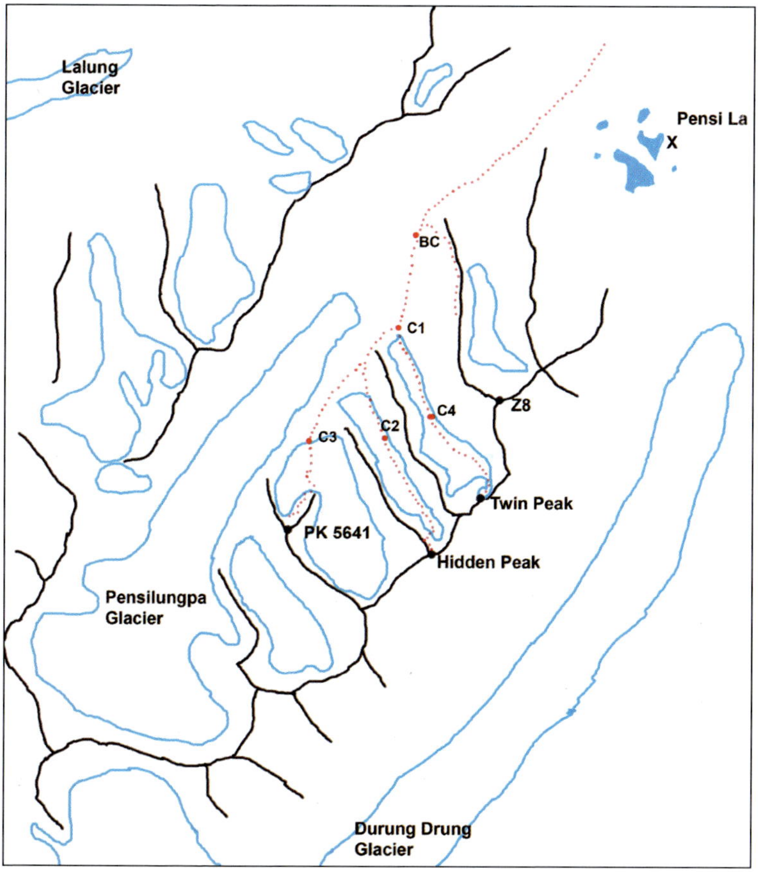

Pensilungpa route and camps.

North face of P. 5641 m. (Derek Buckle)

of the major peaks bordering the Durung Drung glacier, but incoming clouds now concealed the distant views of Nun and Kun to the north. Following our upward tracks we returned to Camp 2 prior to continuing to BC the following day.

After spending a day at BC we established Camp 3 at 4950 m on the true left lateral moraine of the fourth tributary glacier on 16 September in the hope of climbing the dominant 5641 m peak visible from BC.

Route to Twin peak from glacier 2. (Derek Buckle)

Pinnacle on Twin Peak, Hidden peak just visible at half height behind. (Derek Buckle)

This camp was rather lower than we had hoped, but the complex terrain yielded few level sites that were both objectively safe yet had access to clean water. Early the next day all four of us climbed the true left lateral moraine until it was possible to move right onto the rocky north-northeast ridge of P. 5641 m. Easy broken ground then led us to the broad glaciated north face, but being hard ice we soon abandoned this in favour of a return to the rocks. Higher up the difficulties increased as the ridge narrowed and we were again forced onto the ice of the north face until blocked by a prominent rock finger projecting into the face at around 5576 m. At this point we debated the pros and cons of continuing but the way forward was clearly more complex and further than anticipated. The face itself was exposed and reared to over 60° while a return to the rocky ridge looked more technical than we were prepared for. A decision was made to retreat, returning the way we had come. It was a disappointment, but with a higher starting point the outcome may well have been different. The next day we again returned to BC to recuperate.

On 20 September we again left BC to establish Camp 4 at 5186 m close to the true right lateral moraine of tributary glacier 2, although Chris decided to retreat to BC on reaching the site of our earlier Camp 1. Intriguingly, Camp 4 was characterised by numerous small penitents and ice pedestals up to 2 m high capped by large slabby blocks - it was a magical location, commanding extensive views of the upper glacier. As the weather continued to hold good the remaining three of us departed soon after sunrise the next day to follow the corrugated glacier slightly left of its centre to the point where it visibly steepened at the glacial wall. Here we exchanged poles for axes and crampons before continuing as a rope of three. As the slope increased from around 25-30° to nearer

Doda peak (6485 m) and Durung Drung glacier from Twin peak. (Derek Buckle)

40° we trended leftwards to cross several large crevasses rather than expose ourselves to impending seracs before continuing south to a broad col at around 5600 m. Here we veered towards the west along a steep, snowy arête bordered on the left by a line of rocks overlooking the Durung Drung glacier. The arête terminated in two close rocky summits, the second of which constituted an impressively sharp needle rising from a small airy col. We chose to stop at the first, slightly lower summit (5825 m) which we reached in five hours (Alpine AD). Having made the first ascent of the east summit of this peak we appropriately chose to call it Twin Peak. The summit provided panoramic views of the Durung Drung glacier with its surrounding peaks dominated by Z3, long-distance views towards Nun and Kun and, more locally, of the rocky south-west ridge/face of Z8. With few alternative options available we retraced our ascent route back to Camp 4. The next day we rejoined Chris at BC.

It snowed during the night of 22 September and the sky remained heavily overcast most of the following day. With clear weather on the 24, however, Mike and I climbed to 5118 m on the north ridge of Z8 in the company of our liaison officer, Suman Kant ('Happy'), in order to photograph its impressive north face and to observe more closely the first tributary glacier now that we had explored the second, third and fourth. Our route presumably followed a line similar to that of the first ascentionists, but we saw no evidence of their passing.

With our time in Zanskar now at an end we left the Pensilungpa the following day to return first to Ringdom Gompa and from there via Kargil to Leh, Delhi and home.

Participants:

Derek Buckle (leader), Mike Pinney, Chris Storie and Tony Westcott.

Summary

In September – October 2013 four members of the Alpine Club visited the Pensilungpa glacier in the Zanskar region of the Indian Himalaya. In addition to general exploration of the area, they successfully made the first ascents of two summits on the Pensilungpa - Durung Drung divide; P. 5802 m (PD+, tentatively called 'Hidden Peak') and P. 5825 m (AD, tentatively called 'Twin Peak').

BOOK REVIEWS

SHIPTON & TILMAN: THE GREAT DECADE OF HIMALAYAN EXPLORATION. By Jim Perrin. Pp. 424, 2013. (Hutchinson, London, £ 25).

Jim Perrin has a unique writing style. It's difficult to confuse a Perrin book with someone else's. And he chooses his topics carefully: John Menlove Edwards, Don Whillans, and now, Eric Shipton and Bill Tilman. He has been fussing over Shipton and Tilman for many years – just as he did with his previous books – and readers have been patiently waiting.

With unprecedented access to unpublished diaries, letters and journals, and deeply personal, yet amusing conversations with an elderly Bill Tilman, sources which Perrin refers to as 'a random harvest of delight', he presents an unusual story in the genre of mountain exploration: a story of real partnership. Not only partnership in the usual sense of getting the job done, but one that evolves into a unique and productive friendship between two bold, engaging and extremely different characters.

This is the story of the greatest partnership in British exploration history. They lived in that magical era – the golden era – when many of the first western forays into the great ranges took place. In the 1930s Bill Tilman and Eric Shipton pioneered many routes in Africa and the Himalaya and found the key to unlocking Everest. They crossed Africa by bicycle, explored China with Spender and Auden, journeyed down the Oxus river to its source and, with little support, opened up much of the Nepalese Himalaya. In the words of Jim Perrin, 'The journeys of discovery undertaken through two decades by this pair of venturesome ragamuffins are unparalleled in the annals of mountain exploration.'

This book, which arrived on bookstore shelves just in time for the 60th anniversary celebrations of the first ascent of Everest, would logically have Everest front and centre in the story. But remember that Shipton was supposed to lead that expedition before he was 'ousted by the shadiest politicking' and, thankfully, the Perrin looks elsewhere for his major themes. He focuses on the 1930s, when this unique partnership was most productive and inspired. The geographical breadth of their wanderings in the Himalaya and Karakoram is mind-boggling, but even more impressive is their style. As uber-alpinist, Slovenian Marko

Prezelj often says, 'style matters'. It certainly did for these two. Their style not only set them apart from their contemporaries but it inspired future generations of purists. Maybe not 'night naked', as the likes of Erhard Loretan and Voytek Kurtyka managed, but still impressively lean. Decades later, climbers still speak of expeditioning 'Shipton-Tilman style'. High praise indeed!

Jim Perrin is obviously impressed with that style and, as a climber himself, he understands the value and nuances of their minimalism. But his interest in the two goes far beyond that. He digs much deeper into the warp and weft of the fabric of their characters: their skills, their vision, their humour and their pragmatism.

He is fortunate to have the benefit of the vast repertoire of Shipton's and Tilman's writings. As well, both men have been written about extensively and well (although they certainly bear the brunt of Perrin's tongue-in-cheek criticisms in this volume). But Perrin's approach was always going to be different and different it is: he is most interested in the dynamics of the partnership rather than the accomplishments of either one or the other. And in exploring the dynamics of that partnership, inevitably, we learn so much more about the individual characters. In the end, we are flabbergasted that they managed to survive each other. Shipton, the handsome romantic, and Tilman, the silent grump. But the humour! What fun they had together. What fun Jim Perrin had in telling their stories. And how much fun it is for us to relive those moments. Shipton, Tilman or Perrin: it's difficult to say who is the most mischievous.

One of Jim Perrin's strengths is his knowledge of climbing, which gives him a deep understanding of mountain exploration. He is fully aware of this advantage and is not shy about criticising and correcting other authors on these topics, describing them as 'straw-grasping' and 'speculative'. His running commentary with the reader in the footnotes is particularly bitchy, but oh so entertaining. I've heard some grumble about the 'amount of Perrin' in this book. Okay, I agree, maybe he does sometimes insert himself a bit too much, particularly in those footnotes. But it is always entertaining, extremely well written, right on point, and almost always funny. What's not to like?

Not to dwell overly much on humour, this book is extremely well researched, written with that witty, literate Perrin style, and recounted with a genuine passion and respect for his subjects. It shouldn't surprise us that his two main characters, both somewhat subversive in their world, share a clear affinity with Perrin. You may not always agree with his opinions but it's hard not to respect them. This book is destined to become a classic in the genre of mountaineering literature.

If there was one word of warning to a reader picking up this book it would be – don't skip the footnotes. They are an integral part of the book and well worth the effort.

BERNADETTE McDONALD

MY FATHER, FRANK. By Tony Smythe. Pp. 324, 96 b/w photos, 2013. (Baton Wicks, Leicestershire £20).

Mountain writers fall from fashion, inevitably and often unjustly too. Bill Murray's a case in point - reputation lacerated by a particularly acerbic generation of 'Scotch reviewers', to use Byron's terse phrase. It's pleasing to sense that Murray's work is once more being enjoyed and taken seriously. The same may soon be true for Frank Smythe's.

For decades Smythe suffered from the calumnies of T. Graham Brown – his companion on the two great 1920s Brenva Face climbs which launched his mountain career. Brown was godfather to those same Edinburgh reviewers, so Smythe for years has been target of dismissive and jeering comment. This admirable, beautifully written and scrupulously objective biography should change all that and restore Smythe's writing to its proper status as some of the best we possess. It's one of the most enjoyable climbing books I've read in years.

Witty, exhaustively researched and long overdue, it complements and fills out the fine portrait drawn in Harry Calvert's *Smythe's Climbs* of 1985 – a defence of Smythe that appeared at a time when few listened to its intelligent appraisals. I wish better fortune for Tony Smythe's act of filial vindication - an exceptionally accomplished book, balanced, judicious, without axes to grind. It's also – a rarity among present-day mountain-writing, many of the practitioners of which have only the most glancing acquaintance with climbing – written from authentic personal experience. So it grasps the tensions by which disputes arise; is equally understanding and rather gentle about the personal and emotional aspects of his father's life – a remarkable feat considering how adversely these impacted the son.

This balanced magnanimity becomes all the more notable when his book's compared with the recent one by Harriet Tuckey about her Everest oxygen-man father Griffith Pugh, where childhood misery-memoir segues into wholly unjustified displacement of animus onto another father-figure, John Hunt - vilified by Tuckey, unable now to answer slurs which must cause his surviving relatives pain, yet in truth a principled and honorable man of his time, a distinguished soldier and public servant, caught by machiavels and his own ambition in an impossible situation over the leadership of Everest '53.

I learnt a great deal from *My Father, Frank*, not least about one of my own biographical subjects Eric Shipton – another recipient of Tuckey's partisan despite; learnt much too about the politics and practice of 1930s mountaineering from a man who knows his subject by virtue of fingertip acquaintance and empathic wisdom. Tony Smythe must be unique in having published 47 years apart two of the classics in our literature (the first was the evocative *Rock Climbers in Action in Snowdonia* of 1966, illustrated with photos by John Cleare). His latest is a wonderful book. I hope we don't wait 47 years for the next!

JIM PERRIN
First published in *The Great Outdoors* magazine

8000 METRES – CLIMBING THE WORLD'S HIGHEST MOUNTAINS. By Alan Hinkes. Pp 192, 2 maps, 174 photographs, 2013. (Cicerone, Milnthorpe, Cumbria, £ 25).

Alan Hinkes' book should make a fine addition to any library.

Hinkes is the first British mountaineer to have climbed all the fourteen 8000 m peaks. On the Internet, his achievements have sometimes qualified as 'disputed' the ascent of Cho Oyu being a case in point. Hinkes mentions reaching the mountain's vast summit and walking around to ensure that there isn't any higher to go. All this in semi white-out conditions with reduced visibility. Therefore views of other major peaks, useful to establish proof of having reached a summit often remain elusive. Hinkes has also been alone at this stage of the climb. 'I did not bother to take any photos. There was nothing to see and I was more concerned with finding my way back before I became trapped in a full white-out or deadly snowstorm,' he writes in his book. Among the fourteen 8000 m peaks, Cho Oyu is often described as the easiest. Hinkes' chapter on Cho Oyu begins thus: 'Categorising any 8000 m peak as "easy" or referring to an "ordinary" or "normal" route to the summit, is a contradiction in terms. There is nothing easy or normal about any 8000 m mountain. Each of the fourteen giants represents a serious undertaking with different characteristics, dangers, difficulties and local weather patterns, and none should be underestimated.'

In the eyes of the sport's high priests, the situation on Cho Oyu may have inspired lack of precision in summit claimed. But it takes nothing away from Hinkes' book, which strikes a fine balance between a coffee table book and an account of a life in mountaineering, especially that recap of fourteen 8000 m peaks climbed over eighteen years, entailing twenty-seven attempts in all. Towards the end of the book, Hinkes says that the fourteen 8000 m peaks are dangerous, that he climbed them for himself and not for money and therefore even guiding on those peaks for money isn't worth risking his life again. The only mountains from the fourteen that he may consider climbing again are Everest (to which he returned) and Cho Oyu.

The book's biggest strength is the sheer simplicity in his telling of his story. It is unpretentious. This is complemented by large and beautiful photographs backed by an uncluttered layout. The images give you a genuine sense of place without complicated camera work to distract from what is being shown. Hinkes' photography is crisp and clean. The book has a nice architecture in terms of written content. Having devoted the introductory chapter to describing his affection for adventure and the evolution of his career in climbing, Hinkes keeps the accounts of his climbs straightforward and bereft of searching philosophy. He speaks in a matter of fact manner, mostly devoid of the dramatic, adding a touch of drama only where it is relevant. Each story of climbing an 8000 m peak is followed by a smaller chapter on an interesting aside. The latter ranges from the photo of his daughter that he carried to mountain summits (it also gave him something to look forward to after the summit and kept him focused on descending safely), to profiles of Jerzy Kukuczka, Kurt Diemberger and Reinhold Messner, the correct clothing for high altitude mountaineering, his food habits on expeditions, 'the death zone' as extreme high altitude is popularly called and dealing with death in a dangerous sport.

We live in an age, where mountaineering narratives are many. The media gaze has spared no landscape. Some would say - as an early victim of media in adventure, the snowy, windswept heights of our planet suffer from a fatigued idiom of expression. If despite that, if we still indulge the media, then it must be conceded that content matters now more than ever before. Details like perspective, craft and lightness of handling, previously overlooked, have emerged as differentiators for our tired senses. This book, at once serious for the subject it handles and enjoyably light in treatment, lives up to that more comfortable aesthetic.

SHYAM MENON

SHORT REVIEWS: (BY NANDINI PURANDARE)

SACRED MOUNTAIN. By Robert Ferguson. Pp. 342, 2013.
(Matador, Leicester, £7.99).

Sacred Mountain is historical fiction. The jacket summary is as follows: *Philip Armitage was 19 when he was sent to fight in Burma as an army officer. Deep behind enemy lines and cut off from his column, he struggled to save the lives of his loyal Gurkha soldiers by leading them safely back to India. Ten years later, haunted by these memories, he arrives in Kathmandu to report for* The Times *on the 1953 Everest expedition. He finds a city bursting with Tibetan refugees, driven from their homeland by the recent Chinese invasion. Reporters from other papers have also arrived, intent on intercepting The Times exclusive dispatches and scooping the story. Climbing high into the Himalaya, he finds himself the only person able to lead a mission to save one of Tibetan Buddhism's most sacred men, and with him the future of the Tibetan resistance.*

The novel, is therefore an account based on two narratives – the common thread being Armitage, the hero. The story spans Burma in 1943 and Everest in 1953, when the historical first ascent happened. Both narratives create a vivid and historically accurate backdrop for this gripping novel.

Of course, the novelist is ambitious – in one sweep over just 350 pages, he covers the politics and consequences of war, the politics of the Tibetan resistance, climbing Everest and the media dispatches of those times.

The plots are well researched and are full of historical details and written in a pleasing and authentic style. It is a good fast paced read, competent and dramatic.

AMONG SECRET BEAUTIES – A MEMOIR OF MOUNTAINEERING IN NEW ZEALAND AND THE HIMALAYAS. By Brian Wilkins. Paperback, Pp 215, Colour & B/W photos, 2013. (Otago University Press, Dunedin, New Zealand, $45).

The 1950s were very significant for the climbing community in New Zealand. Sir Edmund Hillary climbed Everest and a strong body of world-class New Zealand climbers developed during this decade. It also saw climbers emerging out of the comfort zone of New Zealand's Alps and coming into the Himalaya. This is an important and dramatic book because Brian Wilkins, who was part of the adventure, shares his experiences of these heady times.

Wilkins was closely associated with Hillary. He climbed with Hillary in the Barun valley of Eastern Nepal in 1954 where the latter had a couple of near death experiences. Wilkins describes these accidents and rescue in detail and affectionately etches out the character of this legendary figure at the formative stages in the famous climber's career. Wilkins' New Zealand climbing includes the first ascent of the northeast ridge of Mt Aspiring, a gripping drama of survival as well as a test of the ethics of mountaineering. Although his love and familiarity for the Southern Alps is evident, the romance of the early explorations, of the years when the Himalaya was still relatively uncharted, is where the charm of the book lies. A great effort by a writer who retired as a lecturer in pharmaceutical chemistry!

DICK ISHERWOOD - MOUNTAINEER. An Anthology compiled by John Ashburner. Pp 248, Colour & B/W photos, 2014. (Fast Print Publishing, Peterborough, UK, $56.95).

Richard (Dick) Isherwood, 1943 – 2013, was an iconic mountaineer, who was born in the UK but lived a large part of his life in South east and South Asia, working with *Save the Children*. He was a prolific and superb rock climber who turned his attention to the great ranges in 1964 starting with an expedition to Swat and going on to much climbing in the Karakorams and the Nepal mountains.

On retiring he moved to the US and set up house in the North West where remote mountains and sea kayaking were at his doorstep. He died suddenly leaving a huge void for people who knew and loved him. His brother organised a wake in Cumbria, inviting his climbing partners to participate. Overnight the idea of creating an anthology with contributions from friends, unpublished articles, memos and notes as well as Dick's wonderful articles written for the *Alpine* and *Himalayan Journals*, was born. John Ashburner, a climber and old friend took on the responsibility of putting it all together. He quotes a piece of Dick's writing in the preface of this wonderful collection of articles and essays. It is typical of the capacity that Dick had to paint a vivid picture, not devoid of humour, of a situation that can give anyone cold shivers. The piece refers to an episode during a bivouac on K7 West:

> The boots got passed to me at one end and I hung them all by their laces from a big chock placed in a crack – at least I thought I did. We had a surprisingly warm night, though it was a bit cramped. In the morning I carefully passed the boots back and discovered I'd failed to clip in one of Des's – it had spent the night merely wedged between two others.
>
> I didn't dare confess to this and sat in silence as Geoff said speculatively: 'I wonder what would happen if you dropped your boots down from here?'
>
> Des replied: 'I doubt if you'd get down.'
>
> I spent the rest of the climb wondering just what we would have done.

This anthology has several articles by Dick himself and it traces how his writing style got better and better as his abilities to climb got better. Also, the flavour of the early crazy climbing days as part of the Cambridge University mountaineering club, climbs in Cumbria and Scotland and eventually the big mountains are well recorded by several voices in the book. Hats off to John Ashburner for digging out the little notes and anecdotes, which would not have otherwise seen the light of day.

A note on the publishing – this is a print on demand book which means that a copy is printed only after it is pre ordered.

TIBET: A HISTORY. By Sam van Schaik. Pp 410, Hardcover 2011. (Amaryllis Publishing, New Delhi, Rs. 695/-).

Sam van Schaik brings the history of Tibet to life by telling the stories of the people involved, from the glory days of the Tibetan empire in the seventh century through to the present day. He explores the emergence of Tibetan Buddhism and the rise of the Dalai Lamas, Tibet's entanglement in the 'Great Game' in the early twentieth century, its submission to Chinese Communist rule in the 1950s, and the troubled times of recent decades.

After laying a context of Tibet's extraordinary history, the book talks about the country's complex relationship with China and explains misunderstood aspects of its culture, such its rituals and monasteries and monks, its philosophy and the role of the Dalai Lama.

The author is an expert on the early history of Tibet and Tibetan Buddhism.

This book is highly recommended as an easy to read, straightforward and concise history of Tibet. The best part is that the book tells stories – it does not have the feel of a dull and boring history textbook. And with the geopolitics of this region being what it is, an understanding of the people who have been refugees in India for half a century is well needed.

IN MEMORIAM

ASPI D MODDIE

(1921 – 2014)

'Life is not so much what happens to you, as the way your mind looks at what happens'

Aspi Modie.

Except for a short spell that Billy Badhwar held the post, Aspi could rightly be said to be the first formal Indian member of The Himalayan Club to have been appointed its President. Prior to that the Club's Presidents, as well as it's Honorary Secretaries, were British counting ahead of the time that the remarkable Charles Crawford led the Club from 1948 to 1953 in Calcutta. It was during his tenure that Aspi joined as a member in 1949. Crawford's departure for England marked the virtual end of the dominant British administration of The Himalayan Club and its golden period of influence as the foremost mountaineering entity in this part of the world, with Calcutta as its base and Darjeeling its frontier of operations. Everest was climbed in 1953 and it reflected the gracious period of the Club's style and prominence then. The members of the time to name those who were yet in India and of whom, Aspi used to speak, were Trevor Braham, Bob Lawford, C J Henty, Brian Ritchie, Harold Williams and latterly Soli Mehta who were then, at one time, Calcutta based. Gurdial Singh, Balwant Sandhu, Jack Gibson, Mohan Kohli and later Suman Dubey were to become his friends. Aspi followed Brian Ritchie, the last of the British presence, as President, and held the position with distinction and authority, from 1975 to 1982. Of course everyone in the Club had by then met him or known of him, for at the time of his death at the age of 92, he was the eldest, or near to that the longest and most respected serving member of the Club, covering a period of 65 years.

I had the privilege of knowing him from the time of our youth, though he was nine years senior to me in age. He introduced me to membership of the Club in 1953, though on a trek together in 1950 to the environs of the Satopanth glacier in Garhwal, having taken me under his wing so to speak, he opened for me a world of learning, adventure, poetry, of poets and writers, appreciation of nature, birds and forests, concern for the fragile Himalayan ecology, a belief in the Creator quite different from that of religious texts, and yet he could quote with relative ease from the Vedas and Upanishads, the teachings of Christ and that of Buddhism, extensively from the Koran, an ardent believer in the Vohu Mana, the good mind as pronounced by Zarathustra, with so many writers and sages and their words of wisdom and truth being his constant fellow travellers with Vivekananda, Aurobindo, T S Eliot, Tagore, the Shinto Masters, Einstein, being prominent among them. He read everything that presented itself of value and happening and commented upon it through his numerous articles and writings. His fertile and enquiring mind manifested itself in a non-ending search and whose scholarship covered a universe, including the nitty gritty of economics, sociology, history and environment, that reflected itself in career pursuits. Yet he was down to earth and had a sense of humour that made it easy for him to laugh at himself and he did that often. Disguise it in any way one might, he was, and remained a Renaissance man.

Born to Parsi parents, Aspi grew up in a carefree but highly disciplined home, under the overview of a father who was deeply religious, preoccupied with his career as a civil servant in Bengal and responsibilities of office at the time of the British Raj. The liberality came from the presence of a gentle and uncomplicated mother. The period was also that of the First World War. A portrait of his father in uniform was a reminder of the time, and possibly one which influenced the start of his own career. His father for his contributions to the Raj, was the recipient of an MBE and the title of Khan Bahadur.

Aspi saw boarding school life from the unusual age of four and a half at a distant convent school in Halflong, now in remote NE India hills. When his younger brother Kersy was ready for school at age seven, both were sent off to Victoria, an English boarding school in Darjeeling district, for the rest of their school education. They were then the first two Indian boys there. In his own words, 'a good public school was the best thing our father did for us. It gave us a good education, good masters, the tradition of a House system of Everest climbers (Mallory, Irvine and Kellas), and not the least, competitive toughness and discipline, which most Indian homes, or institutions then did not offer.' In Mallory House, the first seeds dropped of future mountain passions and pre-occupations. In a later capacity and as Vice-Captain of Mallory House, Aspi was the notional 'keeper of Mallory's compass', donated to the school by Mallory's widow. It had been retrieved by Noel Odell at the last camp on Everest of the 1924 expedition at twenty seven thousand four hundred feet. Years later, after Victoria school had undergone a time transformation and was an Indian semi Government school we, in the Eastern Section of the Himalayan Club, made an effort to locate the compass, and for a fleeting moment Chandra Shekar Ghosh (Domo), a member, held the valuable object in his hands, encased in a pouch with the initials GLM. It had remained in the safe of the Bengali Principal of the School who hadn't any idea of what it represented until we explained its history to him. That, and our combined sense of fair play, precluded any attempt to bring the compass with us and make it a part of The Himalayan Club's prized possession and folklore.

As he described it often, it was as a school Scout, that he was introduced first to the beauty and mystery of the Himalaya on a trek to the Singalila ridge, between Darjeeling and Nepal, when from twelve thousand feet at the age of twelve, he gazed in awe on three of the highest peaks of the world, the massive throne of the gods of Kangchenjunga, the distant summits of Everest and Lhotse, and also Chomolhari in distant isolation to the east of Tibet. He caught the mountain infection then and it remained everlasting.

His other passion was History, and again, as he described it, it came from his History Master and Headmaster at Victoria and the urge remained in him through life. They dipped him into ancient Greek and Roman history, also that of Europe and some Indian. The ancient Buddhist mound of the Paharpur monastery near their home in central Bengal stimulated his historical imagination beyond that of books. And so it became an integral part of the man, the intellectual framework as it developed in him with which he viewed his later contemporary world – 'the Indian world post Independence, the Western world after the end of imperialism and decolonisation, the Chinese since Mao, and the Islamic world of fundamentalism, as also the American fluctuation between isolation and economic cum missile strategy.' Whatever were his later wanderings into politics, economics, the sociology of development and environmental science, of which he wrote extensively, it was History that remained the window of his world perspective.

After a sheltered life at school that ended in 1939, two happenings brought him into the real world. The first was going to Allahabad University for a post graduate course in History; the second was joining the Indian Army. He was disappointed in the former as the History Department's reputation as being the foremost of its time exceeded the reality. But Allahabad then was a nursery for all-India services, being the hub of Indian political thought and aspiration, a real Indian environment and that too Hindi-speaking. The shortfall of the learning at Allahabad University was more than made up elsewhere. Anand Bhavan, the home of the Nehru family, was nearby and of easy access in those non-security days. Jawaharlal was a distinguished neighbour, and sight of such people like him, Kripalani, Maulana Azad, Subhash Chandra Bose, Gandhiji himself, brought more understanding of Indian history and the idea of India to young socialist minded college unionists, than all that the University could impart. Nehru's *Discovery of India* had an impact on Aspi's own search and identity with India. Much later, near to the conclusion of his wanderings and career influences, Aspi was to write his literary opuses *The Brahmanical Culture and Modernity; The Failed Mahabharata : Making of the Indian State* and *Leaders of Straw : India's First 50 Years.* The dominant question asked was, did 'Swaraj' mean having British colonial systems and laws in Indian hands, and were there no lessons, positive and negative, learned over the two thousand years of recorded Indian History?

After University, his intention was to sit for the ICS examination, the career of his choice, but British India closed recruitment in 1942 for the duration of the war. With his brother already in the Army and with the Japanese closing in on India's eastern frontiers he too chose to join as an Emergency Commission Officer for the duration of the war. It was of short duration as careers go, with a posting to the Fourteenth Army on the Burma Front. There he experienced chaos and unplanned preparation to what an infantry and staff officer was subjected. He came out wounded in an accident that reduced him to category C, which meant he was unfit for military operations. With a slow process of recovery his days in the army ended with a half job as a member in the air-priorities board in Calcutta, 'a sort of anti-climax of the injured.'

He sat for the ICS Exam in 1945. After his demobilisation in the army in 1947 he was offered a position in the Indian Foreign Service as a consequence of his declared interest. The assignment was that of a Trade Agent to Gyantze in Tibet. He said that he received the offer on the strength of two earlier treks that he had undertaken and a climb on Lama Angden, but family compulsions denied him from accepting it. It was not long before that he had got married and the idea of a posting to a far distant land was not considered right for a young family man. The IAS had been formed and a posting nearer to home by the end of 1948 to Muzaffarpur in the Tirhut division of North Bihar as an Assistant Magistrate became his. Here he came to learn of life in the collectorate, of petitions, litigations and administration, of law as administered in an age old relationship of market place government and those governed at the grassroots in India. He was later posted to Madhubani sub-division in Darbhanga district, on the remote Nepal border, long considered a punishment posting, but it was one that came with independent charge. He described life in 'The Collectorate's versatility, as that, of a court, a bank (with its hub being the Treasury), a market, sometimes a theatre, for the humble peasant's needs that acted as the district headquarters.' As a district Collectorate it provided an example 'how we do it,' serving many purposes. 'We surround dull, austere official business with the atmosphere of *mela* and *tamasha.*' But the roles exposed him to learning the hard realities of life and poverty. In the seven years that encompassed his tenures, he was to witness abject human suffering through shortages of famine, the relief work that compounded it through annual Kosi river floods, of medical needs and lack of provisions, conditions that left an indelible mark upon him for life.

After Madhubani, and a short stint at base in Patna in the textile commission, and a chance meeting led to an opening for him in Lever Brothers, India, the day that his second son was born. It turned out be auspicious for him. He made the change in 1957 based upon new challenges that the world of MNCs and that of professional management offered, of business, budgets, forecasts, planning and markets. So began a new career that extended till 1980, and when it ended he was a director of Unilever India, with his major responsibilities being that of defining business policy and corporate planning. The striking inroads he made in the corporate world meant entry into associated fields, and positions that he held with distinction, as Executive Vice President of

Assocham, faculty member of IIM Ahmedabad, and visiting lecturer of the Indian Institute of Public Administration. He was associated with a number of other national and international organisations at various points of time, and among these he was Member for the Himalayan Region Planning Commission, and Member National Committee of Environment and Planning.

Post retirement, he lived at 'Revills' on Cuffe Parade in Mumbai, and within the leafy environs of his home, he wrote his books and the numerous articles that were published, including those for The *Himalayan Journal* and other publications. The changes he urged were for environmental protection, particularly of the fragile Himalayan region. His one sadness was that he could not enthuse The Himalayan Club to embrace his ideas to introduce effective ecological programmes and despaired on being told that 'that was the role for scientists to pursue and not that of mountaineers.' He brought about changes in The Himalayan Club, not least, that its very continuity, as we have it today, is the direct result of his strong opposition to its merger within the ambit of the Indian Mountaineering Foundation, as some senior members of the club had proposed at the time that its British members were leaving. The trail of mountains and mountain paths flowed throughout his life as he blazed his numerous treks and travels through it. None was as insightful as his own planned reconnaissance of the unexplored route to Everest from the south in Nepal, till it got usurped by the Shipton expedition. He spoke extensively at forums of his travels and was particularly fond of the invitations he received from the Club's Eastern Section and of his association with it.

His great retreat was to 'Savera' his peaceful home in Bhimtal, where he committed most of his ideas to paper. Twice each year he would make his pilgrimage there with his wife. He never tired of telling the tale of the leopard that loped his way over the tinned reinforced roof in a surprise visit. Though a modern man in mind and application, he remained orthodox in that he shunned the computer and the cell phone. He wrote in long hand his manuscripts and letters. The Postmaster at Bhimtal found in him a kindred spirit, as he zealously bought 10p postcards on which to scribble notes to family and friends. His faith in the efficiency of the Indian Postal Service remained unshaken and with it he never lost the art of letter writing. In later life he sought the skills of his granddaughter Mandira, who shared a closeness with him, as she helped him with his manuscripts.

Aspi leaves behind his beloved wife Kitty, son Arvand and daughter Behroz, son in law Riaz, and through them and his late son Varji, seven grandchildren and one great grandchild. His felt a deep loss when his brother and a son died before him.

His final two books were written and published in the last four to five years of his life, that on *Tibet* which he dedicated to the Dalai Lama; and the influential *Witnesses to our Times,* which is also the closest he got to writing an autobiography. He dedicated it to the memory of his beloved son Behram (Varji). In writing this obituary I have quoted extracts from it, though for a detailed insight of his life, one can read it in depth on http://admoddie.blogspot.com (September 2008).

At the start of this obituary there is a quote from Khalil Gibran. It could well have been written by the poet to reflect Aspi, the man, his witnesses and times.

MEHER MEHTA

GEORGE LOWE

(1924 – 2013)

If George Mallory's off-the-cuff justification for climbing Everest, 'because it's there', are the best known words in mountaineering, Edmund Hillary's somewhat less enigmatic summary of the first ascent, 'we knocked the bastard off', must rank a close second.

George Lowe. (© Royal Geographical Society with IBG)

Hillary never intended the remark for public consumption. He delivered the words to his fellow New Zealander and close friend George Lowe some 150 m above the South Col on 29 May 1953. Lowe had climbed from the Col to meet Hillary and Tenzing Norgay as they descended the mountain, but neither he, nor the rest of John Hunt's team waiting anxiously below, knew whether the pair had reached the summit.

'They were moving fairly rapidly - the only tiredness showed in their slightly stiff-legged walking as they cramponed the last part of the couloir,' Lowe recalled. 'I crouched, back against the wind, and poured out the thermos contents as they came. Ed unclipped his mask and grinned a tired greeting, sat on the ice and said in his matter-of-fact way, "Well, George, we knocked the bastard off!" It was not quite matter-of-fact; he was incredulous of what they had done.'

Lowe later felt the need to put the record straight about this apparent profanity. Writing in the 1993 *Alpine Journal*, he said this remark meant no disparagement to Everest and was intended for his ears alone. In New Zealand slang, big mountains were referred to with admiration as 'big bastards', said Lowe. 'Ed had to live with the fact that I told the BBC what he said on meeting me, and he has never forgiven me for it!'

Lowe's sense of fun is in play here. And even were it not, Hillary had little to forgive. Grateful thanks were more in order, for Lowe had played a major part in making a summit bid possible. An accomplished ice climber, he had spent 11 days at around 7000 m, often battered by high winds, pushing the route up the Lhotse Face towards the South Col. Hunt described it as 'an epic achievement of tenacity and skill'.

Then, on 28 May, Lowe had led the way up from the South Col. With Lowe, as well as Hillary and Tenzing, were Alf Gregory and Ang Nyima. At 8340 m they picked up a cache of gear left, after a gruelling carry, by Hunt and Da Namgyal. 'Ed took the tent, Greg the RAF oxygen cylinder and I took fuel and some of Greg's load,' he recollected. By now Hillary was carrying an estimated 63 lbs, Lowe and Gregory 50 lbs each and the two Sherpas 45 lbs each - all way over the 15 lbs reckoned to be a good load at such an altitude. This Herculean effort succeeded in placing a camp at about 8500 m from where Hillary and Tenzing would go for the top.

Wallace George Lowe was born on 15 January 1924 in the farming community of Hastings, North Island, the seventh of eight children, and the seventh child of a seventh child. His father was a fruit grower (an 'orchardist' in New Zealand parlance) and kept bees as a sideline. He got his queen bees from the Hillary family in Auckland - George delighting in the coincidence when he first met Ed Hillary, on a bus heading for the Mount Cook area on South Island where he (Lowe) had got a holiday job escorting tourists to a glacier.

Young George hardly seemed destined to be a climber. Aged nine, he had broken his left arm; it would not mend and had to be realigned seven times. It remained bent and virtually without muscle for the rest of his life. Added to this, he had an early fear of heights. He attributed his embrace of mountaineering to his determination to face up to his weaknesses.

Climbs with Hillary and others in the Southern Alps - notably a first ascent of the long, jagged Maximilian ridge on Mount Elie de Beaumont - led to the Himalaya in 1951 where he and three companions made the first ascent of Mukut Parvat (7242 m) in Garhwal. On the walk out, at the village of Ranikhet, the group picked up their mail, including a surprise telegram from the English mountaineer Eric Shipton inviting two of them to join an expedition about to leave Kathmandu for an exploration of the south side of Everest.

The telegram 'turned four amiable New Zealanders, relaxing in the hill station lounge, into four tense tigers, caged, self-seeking, eying each other with jealousy,' Lowe wrote in his very readable memoir *Because it is There* (London 1959). But he and Ed Cotter were broke and had to watch, 'riddled with envy', as Hillary and Earle Riddiford departed for Nepal.

A year later, however, thanks to Hillary's recommendation, Lowe was invited by Shipton to join the British expedition to Cho Oyu (8201 m), a tough rehearsal for the 1953 attempt on its neighbour, Everest. Cho Oyu remained unclimbed but Lowe and Hillary took consolation in an outstanding first crossing of the torturously crevassed Nup la from which they descended secretly into Tibet. The two colonial boys had certainly earned their ticket to join the chaps next year.

Lowe had trained as a teacher and was working in the same primary school where he had been a pupil. He had been given special leave for his Himalayan trips in 1951 and '52, but a third time was pushing his luck. Leave was refused so he resigned.

Though Lowe would later return to teaching, Everest, almost by accident, opened a second profession, that of photographer and film-maker. Lowe had been a keen 'stills' amateur since his teens and when a bout of pneumonia rendered Tom Stobart, the expedition cameraman, unable to work at altitude, his movie moment arrived. Entering the Khumbu Icefall, it was the first time Lowe had held a ciné camera in his life; action sequences he shot high on the mountain contributed greatly to the success of the expedition film, *The Conquest of Everest*, which was nominated for an Oscar.

These new credentials, together with a familiarity with crevasses, led to his being invited by the geologist/explorer Vivian Fuchs to join the Commonwealth Trans-Antarctic Expedition — an enterprise that occupied Lowe for three years. In a tracked Weasel vehicle called *Wrack and Ruin* he made the 2158-mile crossing in 99 arduous days, uniting with Hillary at the South Pole on 19 January 1958, his fellow Kiwi having approached from Scott Base in the opposite direction. Between Everest and Antarctica, he joined Hillary for an attempt on Makalu (8481 m) that came close to disaster when Hillary cracked three ribs in a crevasse rescue and became ill, and later in a hunt for the yeti (not found).

In 1959 Lowe returned to education, first at Repton School, Derbyshire, and then, from 1963, at the Grange School in Santiago, Chile, where he became headmaster. In 1962 he had married Hunt's daughter Susan and their three sons were born in Santiago. The family returned to the UK in 1973 - the year of the military coup in Chile and the death of Salvador Allende - after which Lowe worked as one of Her Majesty's Inspectors of Schools until retirement in 1984.

Lowe, like Hillary, felt a deep commitment to the Sherpas of Nepal and in 1989 he helped found the UK arm of Hillary's Himalayan Trust, funding schools, health services, and other projects. The work was shared by his second wife, Mary, who continues as the Trust's secretary from their Derbyshire home.

Lowe has been described as the 'forgotten hero' of Everest. It is unlikely he minded. Though witty and amusing company, Lowe shunned the limelight and expressed relief that he was not in the summit party. 'Ed

Hillary was the right one. I wouldn't have had the diplomacy that he had,' he said. Lowe's part though is likely to get a reassessment in the coming weeks with the publication of two books: *Letters from Everest* (Silverbear) and *The Conquest of Everest* (Thames and Hudson), both being a collaboration between Lowe and the historian Huw Lewis-Jones.

Lowe died in Derberyshire on 20 March 2013. He was the last surviving member of the 1953 Everest climbers. When the 60th anniversary of the first ascent is celebrated in May, only the author Jan Morris, who was *The Times* reporter attached to the expedition, will be able to recall the moment when Hillary and Tenzing descended to Hunt's advance base camp in the Western Cwm and George Lowe, leading the group down, gave an exuberant thumbs-up. Soon the world knew that Everest had been climbed.

STEPHEN GOODWIN

(This obituary was first published in *The Independent* on 26 March 2013).

IAN McNAUGHT-DAVIS

(1929-2014)

Ian McNaught-Davis was a computer specialist, television personality and climber held in high respect among the international mountaineering community. In 1956, with his climbing partner Joe Brown, he made the first ascent of the Muztagh Tower, a 7000 ft sheer wall in the Karakoram, reputed to be impossible. As a skilful negotiator he spent 10 years as president of the International Mountaineering and Climbing Federation (UIAA), when his energy and incorrigibly good humour guided the Federation

Ian McNaught-Davis.

through a difficult period of growth and change. His quick wit and beguiling charm opened a parallel career as a commentator and droll observer of the climbing scene, in particular such television epics as an ascent of the Old Man of Hoy in the Orkney Islands, in which he also took part, and a series of eight adventure films by Leo Dickinson. They covered a range of daring topics, from canoeing down the so called 'River of Everest' to reaching the mountain's summit without oxygen. McNaught-Davis's unsensational Yorkshire tone allowed the images to speak for themselves, adding to the drama. Similar forays into television saw McNaught-Davis climbing the Eiffel Tower in Paris with Lionel Terray and questioning aspiring climbers at the foot of the Eiger North Wall about why they should want to climb it. Success in this part of his career earned him the title of 'Mac the Telly'.

McNaught-Davis was born in Wakefield and won a scholarship to Rothwell Grammar School. He spent two years on National Service in the RAF, hoping to follow the example of his father, Stanley, who was a pilot throughout the First World War before helping to establish probably the earliest air route between London and Warsaw. Ian was enthralled by his father's stories of flying sorties over the Western Front and later as a test pilot for the Polish Air Force, stories illustrated by photographs of Stanley standing alongside an impressive number of crashed aircraft. Unfortunately, defective eyesight ruled out any hope of a flying career in the RAF for Ian who served his two years as an aircraftman clerk. After demob McNaught-Davis began studies at Manchester University, graduating in mathematics. At university he also began his life-long love of mountaineering at which he showed immediate talent, although his appearance as a bespectacled rugby forward lacked the fashionable athleticism of later generations of climbers. He was an immensely strong and determined mountaineer who made a number of first ascents on British cliffs, in Greenland on the Hjornespitze and Berscarkerspitze and in the Russian Pamirs and on a British-Russian exchange expedition.

McNaught-Davis was less than impressed by the authoritarian attitude of the Russian hosts towards mountaineering. His habit was to cheerfully greet Vladimir, his host: 'well done, Bloodymir! You mad, impetuous Russian dare-devil!' Strict Soviet rules and sense of etiquette made the relaxed attitude of the British climbers bewildering to their Russian hosts. On the summit of the Peak of Communism, the

Russian leader, observing McNaught-Davis lighting a celebratory fag, demanded to know: 'How is it Mac that you climb so well when you are so decadent?' The reply was not recorded but the difference in attitude was palpable.

McNaught-Davis was at the forefront of post-war university climbers, making first or early guideless British ascents of such difficult routes as the Pear Buttress of Mont Blanc and the North Wall of the Cima Grande in the Dolomites. Apparent recklessness as a car driver or motorcyclist was well known and as a raconteur he had few equals, but as a mountaineer, bulldog strength was matched with canny judgement. Joe Brown remembers him as someone slightly larger than life, a complete individualist capable of mixing successfully with any level in society without intruding, a sort of citizen of the world.

As a mathematician the advent of computers in the early 1980s gave McNaught-Davis a sharp new interest and he entered the industry, working for BP in oil exploration in East Africa and Singapore before returning to Britain as managing director of the British subsidiary of the American company Comshare Inc. handling the company's European business. His fascination with computer technology and grasp of its potential, in days long before the birth of the World Wide Web, was strikingly prescient. A chance meeting with a television contact in a London pub resulted in an outline for a series of programmes explaining, in layman's terms, how computers worked and what their potential was for disseminating information. This in turn led to a 30-part series on BBC television with Chris Serle and McNaught--Davis explaining the mysteries of computer technology. His easy style and ability to explain complex ideas in simple terms made the series a major success. 'Mac the Telly' had indeed mastered the medium.

McNaught-Davis became a patron of the British Mountaineering Council and was a former president. He continued climbing into his 80s and Simon, one of his two sons, recalls sharing many adventures with his father. On one steep climb in the Canadian Rockies, a fall almost pulled both of them off the mountain. They survived and safely descended to the nearest town only to be told to move on because a Chapter of Hell's Angels was about to appear and mayhem to ensue. McNaught-Davis was fascinated and immediately sought out the leading Angel, a fearsomely tattooed individual, disarming him with his thoughts on the relative merits of British and American motor cycles. They ended bosom friends, Simon recalled. He was just that sort of chap.

Three years ago McNaught-Davis was diagnosed with Alzheimer's disease and nine months ago with liver cancer. He is survived by his first wife Mary Alderman from whom he divorced and his second wife, Loreto. He had two sons Simon and John with his first wife and a daughter, Sarah, who died through illness in Africa.

Ian McNaught-Davis born August 30, 1929, died February 8, 2014 aged 85.

RONALD FAUX
(Published earlier in *The Times*)

Ian McNaught-Davis was an honorary member of the Himalayan Club and was instrumental in coordinating and supporting the Club's activities in UK and Europe. – Ed.

RICHARD (DICK) ISHERWOOD

(1943-2013)

Dick Isherwood's first expedition to the mountains of Asia was to Swat Kohistan in 1964, and his last trek to Manaslu was in late 2012, just a couple of months before his sudden death. Over nearly half a century in between he worked, climbed and travelled all over the Himalayan region.

Dick Isherwood. (Geoffrey Cohen)

Dick was born in Lancashire in the north of England and educated at Manchester Grammar School and Cambridge University, where he read zoology and did a postgraduate course in agronomy. He was a leading rock climber on the English and Welsh scene of the late 1960's, accomplishing many of the hardest routes of that time. However he moved to Bangkok in 1969 and stayed in Asia for the next fifteen years. Subsequently he moved to the

USA, but continued to make frequent visits to Asia for both business and pleasure.

Among Dick's most significant first ascents were P. 6127 m in South Parvati with Rob Collister in 1973, Lamjung in 1974, Kanjiroba in 1976 and Dorje Lakpa in 1979. Additionally he made a number of early re-ascents or attempts on peaks in the Karakoram/Hindu Raj (e.g. Thui II, Buni Zom, K7 West, Drifika); Nepal (e.g. Annapurna II, Chulu East, Bauddha, Tsoboje) and China (e.g. Haizi Shan, Yangmolong). He climbed widely beyond the Himalaya too – making an early ascent of the Carstenze Pyramid in 1972, and ascents of many peaks in the northwest USA after he moved there in 2000. But perhaps what he enjoyed as much as the climbing expeditions was trekking through interesting and often unfamiliar territory. He was a skilled naturalist, with a particular interest in birds[1] and was rarely to be seen without a large pair of binoculars.

Dick's willingness to carry huge loads and tirelessly make steps in deep snow naturally made him a force on any expedition. Equally or perhaps more important he had very good strategic mountain sense, so his choice of where to camp, how much time to allow, when to start and when to turn back were infused with his ability to rapidly appraise the topography and climbing potential on a chosen objective. On the marches to and from the mountain he always relished the local food and drink, becoming something of a connoisseur of *chang* and *rakshi*.

In 1977 Dick took up a post to manage a UK-aided farm at Pakhribas in eastern Nepal. This gave him a hugely enjoyable two years in which he enlarged his knowledge of Nepali and did a vast amount of solo and accompanied trekking. He seemed to understand the rhythms and exigencies of village life in the Nepalese hills, and was at home with its variety of people: Tamangs, Gurungs, Gurkhas, Limbu and Rai. The agricultural project developed successfully under his leadership; although he sometimes expressed frustration with the bureaucracies involved in overseas aid, he had a real feeling for what would and would not work. At the end of two years, needing a change, he tried his hand at leading treks for the burgeoning American trekking company Mountain Travel. He was a popular and highly competent trek leader but Dick was too interested in exploring byways for this to become a career choice.

1. *HJ* Vol. XXXIV (1974-5), p. 130. 'Birds of Swat and Gilgit'.

He spent about 18 months cycling around India's bird sanctuaries and archaeological heritage sites in between a variety of mountaineering sorties[2], then, after a year back in UK, he returned to work for *Save the Children*, first in Bangladesh and then in Nepal. With his wife Janet they adopted a child in each of these countries, then after a spell in Singapore they moved to the eastern USA to provide a steadier base for raising a family. Dick's work still took him regularly all over the world supervising agricultural field research trials, but his family responsibilities allowed less opportunity for climbing expeditions. They lived next to a small lake and having taught himself to roll a kayak he took up sea kayaking on the east coast of the US, allowing him to pursue his bird-watching interests while seeking out remote places nearer home.

Aged 55 Dick retired and relocated to Port Townsend on the Olympic Peninsula of northwest USA, where remote mountain and sea kayaking trips were on the doorstep. He took to making winter canoeing journeys in Baja, California and made a fascinating canoeing trip to Bylot Island in the Canadian Arctic. Later he acquired a sailing boat and indulged in exploration of the San Juans and other nearby islands. He also returned many times to the Himalaya. In 2000 he joined a party making a winter traverse of the frozen Zanskar river; in 2002 he was on an Alpine Club expedition to Pokharkan in Nepal and between 2004 and 2009 he made four expeditions to Sichuan; in 2010 he joined us for an expedition to Sikkim. There were also frequent treks in Nepal, often with his son Sam.

Dick had a wonderfully dry sense of humour, employed to great effect in his many articles for the *Alpine* and *Climbers Club* Journals. In 2005 he took on the task of compiling the Nepal section of 'Area Notes' in the *Alpine Journal*, the annual summary of principal mountaineering achievements. In 2009 he added the 'Pakistan Notes' to his portfolio. With his vast experience of exploratory trekking in both countries, and ability to quickly grasp the essentials of complex mountain terrain, he was an ideal person for this task.

In his later years Dick sported a huge white beard. His grandfatherly air combined with some pidgin Nepali led to genuine camaraderie with porters and local villagers, who sensed his real sympathy with them. There was an occasion in the Alps when we arrived at a hut below

2. *AJ* 2010-11, p. 168. ' A Year off Work'.

Monte Viso where some Nepali Sherpas were doing summer work. These young lads were amazed to be hailed in their native tongue by this Father Christmas-like figure. Hours later as the rest of us went to bed the Nepalis were still grouped around Dick in a corner having a great blether and hanging on his every word.

Dick died unexpectedly of an intestinal haemorrhage in February 2013, when he might have expected to have many more years of exploring the more remote corners of the earth. He had many friends in UK, USA, Canada and Asia and we shall all greatly miss him.

GEOFFREY COHEN

R MARTIN SCOTT

(1941- 2013)

Having a theoretical physicist as a father and a classicist as a mother, Martin grew up in an intensely intellectual environment. While he undoubtedly acquired his considerable intellectual prowess from his parents, he chose not to follow an academic career, but to plough his own furrow in industry after completing his degree in physics at Merton College, Oxford. On graduating, he first joined Schlumberger as a geophysical engineer and subsequently spent several years in exotic desert locations prospecting for oil. Feeling the isolation of this work he then joined ICL (International Computers Limited) where he rose to become President, in France, of what was now Fujitsu-ICL after twice turning round ailing sections of the company. It was during his long spell in Belgium and France that he became fluent in French.

Martin Scott. (Al Scott)

Martin's ability and strong competitive talents became evident at a very early age when, at school in Cambridge, he became both Head of

School and swimming captain. At school he broke the record for the 100 yards breaststroke in a time that stood for many years. He retained a lifelong enthusiasm for swimming and regularly took to the local waters in Hampstead Pond. Martin also had a long-standing love of mountains, developing particular affinity for Scotland and subsequently elsewhere. His first recorded exploits were in the Zillertal around 1957, with more challenging ascents in the Picos de Europa in 1962, just prior to his long sojourn overseas. On returning to the UK he was persuaded to join the MAM (the Midland Association of Mountaineers) and later the Alpine Club in 1974. He was an active participant in both clubs, becoming a regular attendee at family meets after his marriage to Julia in 1973.

As a long-standing, active member of the Alpine Club, Martin was elected to the committee in 2001 and went on to serve both as Honorary Secretary (from 2003) and as Vice-President in 2008. Like all tasks, he took these positions seriously and put considerable energy into changing the AC into an organisation more fitting of the 21st Century. In this he initiated the electronic bulletin as a more versatile means of communicating with members and was a major supporter of the AC Climbing Fund. He also represented the Club as Chairman of the Mount Everest Foundation and as its representative at the UIAA.

I first really got to know Martin in 2002 when the AC organised an expedition to the Lemon Mountains of Greenland. Partnered together we made six first ascents and three new routes on previously climbed mountains as well as completing a significant circuit on skis. Since then we regularly climbed together, in Europe and in Asia. Particularly memorable were three expeditions to Tibet - to the Nyenchentanghla range and western Tibet - and another to the Obra valley in the Indian Himalaya. On every occasion we successfully made one or more first ascents while exploring the unfamiliar terrain. It was a productive time and Martin's fertile mind and broad range of intellectual interests rarely let the lively evening discussions dry up.

In Europe Martin was an active participant at the Club's 150th Anniversary in Zermatt in 2007 and he also took part in several of the 150th anniversaries of first ascents made by the early AC members. Notable among these was a re-enactment of the first ascent of La Grande Casse by Vanoise Guides in 2010. Fluent in French, Martin maintained a lifelong friendship with many well-known French climbers in the Provence region where he and his wife Julia own an apartment. It was a great pleasure to join him in this predominantly sunny part of France where extensive rock climbing at a variety of levels is available on Mont St Victoire, Chateauvert and, of course, the iconic sea cliffs of the Calanques. Being in France the evenings shared with his many French friends over one or more bottles of fine wine were especially memorable!

Latterly Martin migrated seamlessly into cycling, a pastime in which he also had a long-standing interest. He was an active member of the Aix Cycling Club that regularly organised energetic day tours. I vividly remember joining one of these events during which, being a non-cyclist, there was considerable concern that I should not overdo it. Martin, of course, did it all in his stride and went on to complete several long-distance multi-day tours supported by Julia. Especially notable were his traverse of France from the Channel to the Mediterranean and completion of the gruelling L'Étape du Tour on the route of the Tour de France up the Col du Galibier and the Alp d'Huez on his 70th birthday.

When, late in 2012, Martin was diagnosed with disseminated kidney cancer it was a very sad day for all those that knew him for his integrity, generosity and companionship. He died relatively peacefully on 31 May 2013 surrounded by his family.

DEREK BUCKLE

(Reprinted with the kind permission of the Editor, *Alpine Journal* 2012.)

Martin Scott was also actively associated with the Himalayan Club and for the last few years was hon. local secretary for the UK. In this capacity he revitalised UK membership and also convinced many young mountain enthusiasts to become members of the HC. He also ably represented the club on many occasions within the UK and European mountain fraternity. He with his wife Julia made it a point to attend the 80th Year celebrations in Delhi and joined other HC members for a trek in Garhwal. His passing away is a great loss for the Himalayan Club. He will be remembered fondly for his enthusiasm, wit and astute observations.

(Ed.)

NAWANG TOPGAY SHERPA

(1930-2012)

Nawang Topgay.

Nawang Topgay Sherpa passed away on 02 November 2012. Born in the Tibetan year of the Ram, he was 82.

In April 2012, our team of three, in Darjeeling to talk to and record the lives of the climbing Sherpas, had walked down some precarious stone steps to a broken down hut. In a room no bigger than an alcove, Topgay lay in a narrow cot. His arm which had broken some time earlier and had not healed properly lay curled in an awkward position on his chest. He responded to our presence with a grinning 'Namaste' and when his daughter-in-law, Nima, asked whether he understood what we were saying, he replied 'Bhujo, bhujo' - I understand - in Nepali. The warm and smiling Nima and Da Norbu, his son, told us of his life over tea and biscuits. They showed us the carefully preserved medals and citations. As a result of our visit, the Himalayan Club started sending a monthly stipend to Topgay to help cover the cost of his medicines and a few things besides.

Dorjee Lhatoo reminiscences as a boy, living next door to the energetic, teenaged Topgay in Toong Soong basti in Darjeeling in the early 1950's - those were the days he says, when many Sherpas continued coming to Darjeeling from Solu-Khumbu in Nepal, to find work with mountaineering expeditions and when many history-making first ascents of the great Himalayan peaks were being achieved.

Young Topgay too had migrated in search of work. He was taken under the care and guidance of his father's younger brother, Tenzing Norgay and lived with him till he found a place of his own. It was with Tenzing that he began his expedition work as a high altitude Sherpa carrier.

Those were the times of the outstanding and famous Sherpa Sirdars such as Ang Tharkay, Pasang Dawa Lama, Ang Tsering, Tenzing

Bhote or Norgay, Ajiwa, Wangdi Norbu and many others still doing expedition work.

When Tenzing was appointed Sirdar on the '53 Everest Expedition, he took on his two young nephews: Topgay and Nawang Gombu, his sister's son. They both performed outstandingly well and in recognition of their achievement, they, with other select Sherpas were awarded the Himalayan Club Tiger Medal. The climb also coincided with the coronation of Queen Elizabeth II, and with other Sahibs and Sherpas, they received the Coronation Medal too.

A year after Everest, the Himalayan Mountaineering Institute was founded in Darjeeling with the objective of introducing and promoting the sport of mountaineering among the youth of India. Seven outstanding Sherpas were selected to form the cadre of instructors under Tenzing Norgay as Training Director and Topgay was one of them.

He was later selected to teach at the Nehru Institute of Mountaineering, Uttarkashi when it was founded in 1965. Known popularly to scores of students as 'Guruji' (teacher), his cherry cheeked youthful face can be seen in many early pictures of those times. Guruji retired from NIM and returned to Darjeeling where he spent the rest of his life.

We visited Topgay again on our second visit to Darjeeling in October. The setting this time around was much better. In a new apartment built above his daughter-in-law's family home, Guruji lay in a cheerful room with a view of the valley below

Himalayan Club wreath made by P.T. Sherpa.

Toong Soong visible through his open window. He barely acknowledged us this time, muttering incoherently and vehemently. His loving family was still around him, raising him up, plumping his pillows and keeping him clean and comfortable. A few days later he died quietly and without any fuss. The entire community gathered to send him off in a traditional Sherpa funeral with his body curled up in foetal form, leaving the world just as he entered it. Nawang Topgay was the last surviving Sherpa Tiger medalist in Darjeeling and his passing brings an end to a historic chapter in the story of Himalayan mountaineering. He is survived by his two daughters and two sons and their families. Nawang Topgay, RIP.

DEEPA BALSAVAR AND NANDINI PURANDARE

The Sherpa Project
(Based on information by Dorjee Lhatoo)

THE HIMALAYAN CLUB NEWSLETTER
No. 66 (2013 – 2014)

THE HIMALAYAN CLUB COMMITTEE 2014

(Founded on 17th February 1928)

Leh	Motup Chewang
Manali	Mahavir Thakur
Mussoorie	Krishnan Kutty
Mumbai	Rajendra Wani
Pune	Dr. Raghunath Godbole
Shimla	Deepak Sanan

Overseas

Australia	Garry Weare
France	Claude Gardien
Japan	Yoshio Ogata
Korea	Bae Seung Youl
New Zealand	John Nankervis
Pakistan	Nazir Sabir
Spain	Jose Paytubi
South Africa	Dr. S. A Craven
Sweden	Ake Nilsson
Switzerland	Eric Bernhardt
U.S.A.	Donald Goodman
	Nicholas Clinch
	Paddy Iyer

Our very active HLS in the UK, Martin Scott had passed away last year. The Himalayan Club is looking for an able and active member to represent it in the UK.

Hon. Editors Rajesh Gadgil, Nandini Purandare

Hon. Librarian Grp Capt. V. K. Sashindran

Hon. Asst. Librarians

Mumbai	Rajeev Das
Kolkata	Debraj Datta

Hon. Equipment Officer Rajendra Wani

Hon. Asst. Equipment Officer

Kolkata	Subhashis Roy

Hon. Asst. Treasurer

Kolkata Rajarshi Roy

Web Master Rajan Rikame

E-Letter Editor Dr. Ravi Mariwala

E-Group Moderator Cdr. K. B. Singh, Divyesh Muni

Patrons of The Himalayan Club

Chief Of Army Staff
Anil Ambani
Niraj Bajaj
Adi Godrej
Jamsheyd Godrej
Naveen Jindal
Anand Mahindra
Vijay Mallya
Pawan Munjal
Kunj Trivedi

President Emeritus Dr. M. S. Gill

Editor Emeritus Harish Kapadia

Honorary Members of The Himalayan Club

Bill Aitken
Aamir Ali
Eric Bernhardt
Sir Christian Bonington, C.B.E.
T. H. Braham
Kurt Diemberger
Lhatoo Dorjee
Lindsay Griffin
Kamal Kumar Guha
Thomas Hornbein
Sadashige Inada
Capt. M. S. Kohli
Yuichi Matsuda
Bernadette McDonald
Tamotsu Nakamura
Dr. Tsering Norbu
Jozef Nyka
Yoshio Ogata
Hiroyoshi Otsuka
Robert Pettigrew
Gulab Ramchandani
Mark Richey
George Schaller
Doug Scott
Gurdial Singh
Stephen Venables

Expeditions to the Indian Himalaya 2013

In the year 2013, there were 31 expeditions from foreign countries and 67 Indian expeditions to the Indian Himalaya. Though majority of them aimed for the peaks routinely climbed, few attempted new peaks or routes. Unfortunately, many pre-monsoon expeditions to Garhwal and Kumaun suffered due to unprecedented heavy rains and after effects thereof. These floods in the upper reaches of the Kedarnath valley trapped more than 50,000 pilgrims lower down. It is reported that at least 15,000 pilgrim died. The area has not yet returned to normal.

Select expeditions to new areas, new peaks and routes are mentioned below.

ARUNACHAL PRADESH

Chaukan Pass (2419 m)

Chaukan pass, located on the India-Burma border was reached by team from Mumbai, led by Harish Kapadia. Trekking on the muddy Miao-Vijaynagar road (157 km) they reached the Noa Dihing river. For several crossings of the river they had to construct small bridges and cut across thick forest. Finally a steep climb led them to the pass, not visited by a civilian party for decades. During the World War II a British party crossing this pass from Putao to Miao was trapped due to monsoon floods river and barely managed to survive.

The team also crossed into Burma from Pangsu pass and viewed the historic 'Lake of No Return'.

(Article *HJ* 69)

SIKKIM

Tingchenkhang (6010 m)

Expedition / Organisers: Assam Mountaineering Association, Guwahati

Leader and (Members): Manash Brooah (15)

Period: May 2013

After establishing base camp near Lamuney, the team established one high camp on the western slopes of the mountain. On 14 May, the team attempted to reach the summit but had to stop at about 5580 m as the ropes fixed earlier were either damaged or were carried away by an avalanche. One more attempt was made on 15 May but again the summit proved elusive.

UTTARAKHAND

Trisul (7120 m)

An Indian army team, under the leadership of Brigadier Ashok Abbey (President, The Himalayan Club), approached this high mountain from the Raunthi glacier. They climbed the west ridge of and established three high camps. On 30 May 2013, a team comprising of Naib Subedar Madan Singh, Hav Matwar Singh, Lance Naik Min Bahadur Tamang, Rifleman Yashpal and Paratrooper Ashok Kumar left Camp 3 at 02.30 a.m. and after crossing the col between Trisul I and II reached the summit at 09.20 a.m. The team also had planned to traverse all the three summits of Trisul massif but due to unstable snow conditions, the traverse was called off. On 31 May, Nk Bishnu Bahadur developed major symptoms of high altitude sickness at Camp 3. An excellent team work resulted in a successful evacuation and he was safely brought down to base camp by his fellow members. On 03 June, another team comprising of Subedar Kunwar Singh, Capt Brijesh Kumar, Rifleman Pratap Singh and Rifleman Sunil Rai also reached the summit of Trisul I.

(Article *HJ* 69)

Kuchela Dhura (6250 m)

In the post monsoon season of 2013, an expedition to this unclimbed peak was sponsored by Indian Mountaineering Foundation and led by Dhruv Joshi. After crossing Martoli village, base camp was established at 4173 m on the right bank of Lwan gad. They placed two higher camps and the summit camp was placed at 6064 m. On 6 September 2013, the leader with Vijay, Chitramohan and Wallambok reached the summit and achieved the first ascent of Kuchela Dhura.

(Note *HJ* 69)

Nanda Devi East (7434 m)

In early May 2013, the team under the leadership of Anindya Mukherjee, established base camp at Bidalgwar (4300 m) and fixed ropes to Longstaff col. After establishing Camp 2 on the narrow south ridge at 6200 m, the team attempted to reach the summit in semi-alpine style but had to abort the climb due to high winds and technical difficulties.

(Note *HJ* 69)

Manda II (6568 m)

Expedition / Organisers: Nilkantha Abhijatri Sangh, WB.

Leader and (Members): Shekhar Ghosh (10)

Period: May June 2013

The team established base camp on 30 May on the right bank of Kedar ganga river opposite Kedar Kharak at 4400 m. They established three higher camps on the western side of the mountain. Their fixed ropes exhausted after reaching at 6120 m. They decided the climb further without fixed ropes was risky and hence decided to call off the attempt 9 June 2013. On 11 June one of the members, Biswajit Bose, suffered respiratory problems at base camp. The team evacuated him and took him to Gangotri on 13 June but at the local health centre he was declared dead.

Mana (7274 m)

Expedition / Organisers: Indian

Leader and (Members): Cyrus Shroff (4)

Period: May 2013

The team established base camp (4120 m) at the confluence of Uttar and Dakkhni Nakthani gad on the right bank of the river on 18 May. The team entered the Uttar Nakthani glacier and established three high camps, fixing few ropes on difficult portions. First they tried to reach Gupta khal but then shifted to a gully leading to the upper icefields. They established their highest camp (Camp 2) at 5600 m on 26 May. Unfortunately, the weather turned bad and gale force winds made movement dangerous and tricky. Therefore, after reaching 5900 m, they decided to abandon the climb.

Chaukhamba IV (6854 m)

Expedition / Organisers: American

Leader and (Members): James Ryrie Norton (2)

Period: September 2013

Brief Details: On 15 September, the team established base camp at Sundervan in the Gangotri glacier and attempted the mountain from the western / southwestern side. After ABC, they established Camp 1 at 5180 m but had to call of the attempt because of heavy snow fall in the area.

First Crossing of Chaukhamba col (6053 m)

In July 1912, C. F. Meade reached the col between Chaukhamba I and Januhut, hiking from a high camp on Bhagirath Kharak glacier. He did not cross the col. He stood there for a while, got a glimpse of the Gangotri glacier on the other side, and then returned the way he came. In May 2013, Debabrata Mukherjee and his team of four climbers and two high altitude supporters climbed up the Bhagirath Kharak glacier to the Chaukhamba col and descended to the Gangotri glacier, thus linking Badrinath with Gangotri.

The team was awarded the Second Jagdish C Nanavati Award for Excellence in Mountaineering.

(Article *HJ* 69)

HIMACHAL PRADESH

Khhang Shiling (6360 m)

Cosmin Andron and his friends from Romania joined 'Climbathon 2013' organised by the Indian Mountaineering Foundation in the Bara Shigri glacier in July August 2013. During the event Cosmin with eight others including his friends and a few trainees climbed Khhang Shilling (6360 m) by a new route from northwest face from a high camp on Bara Shigri glacier. They also attempted Kullu Makalu (6349 m) but had to turn back due to an accident that occurred due to rock fall.

(Note *HJ* 69)

KR - IV (6340 m) and V (6258 m)

1. Expedition / Organisers: Coalfield Nature Lovers, WB.

Leader and (Members): Paresh Chandra Nath (5)

Period: June 2013

On 12 June, the team established base camp (4700 m) near Baralacha la. After establishing two more camps, on 19 June, leader with Subrata Roy and two Sherpas reached the summit of KR – V.

2. Expedition / Organisers: Bhadrakali Pathfinders Adventurers, WB

Leader and (Members): Debabrata Mukherjee (9)

Period: September 2013

On 30 August the team began approach march from Baralacha la towards Chandratal and reached Transit Camp (4718 m) and on 31 August their base camp (4955 m) beyond Topko Gongma. After continuing on the Koa Rong glacier skirting below the hanging glaciers of KR – VII they established Camp 1 (5304 m) on 3 November. The summit camp was established on 5 November at 5705 m below the east face of KR – V. On 6 September, by 3.15 p.m., they had climbed up to 6089 m by a new route on the east face towards the summit but had to turn back because of the lateness of the hour and lack of proper protection. They feel that most of the earlier teams have not identified KR – V properly and thus have made wrong claims. (The HC mountain climbing authentication team led by Divyesh Muni prima facie agrees with this statement, their detailed analysis will be published in *HJ* 70.)

KR – II (6187 m)

Expedition / Organisers: Kolkata Trekkers Youth, WB.

Leader and (Members): Spandan Kumar Malik (7)

Period: August September 2013

After establishing base camp at 4650 m, they established three high camps and climbed the east ridge of the mountain. From the summit camp at 5630 m, on 30 August, Mohene and Indranil with three high altitude supporters reached the summit at 7.50 a.m. in whiteout conditions. Though they did not have any clear views, they could locate and photograph the snow picket left by a team from West Bengal (Sonarpur Arohi Club) in 2011 at the summit.

P. 6080 m, Miyar valley

Expedition / Organisers: British

Leader and (Members): Andrew Nisbet (4)

Period: June 2013

In 2012 the Scottish Zanskar Expedition attempted a couple of unclimbed 6000 m peaks in the Namkha Tokpo. They had climbed one peak and attempted P. 6080 m. However, a potential route up the south face was spotted, and in June 2013 Rob Adams, Bob Hamilton, Steve Kennedy, Andy Nisbet and Susan Jenson returned to attempt an ascent from the Himachal Pradesh (as opposed to Zanskar) side of the range. After establishing base camp (4500 m) and ABC (5400 m) in the Miyar valley, all the team members climbed the southwest face of the mountain to reach the summit on 20 June. Many smaller summits were also attempted by them.

JAMMU & KASHMIR

Zanskar and Kishtwar

Hagshu (6515 m)

In July 2013, a team consisting of Bryan Hylenski, Jake Preston, Jonn Jeannerret, Gabe Thomas and Dan Kopperud aimed for Haghshu, P. 6055 m and P. 6035 m located in the Hagshu glacier. They approached the area starting from Akshow village near Ringdom and placed three high camps above ABC. They could reach 6440 m on Haghshu before turning back due to bad weather conditions. They achieved two first ascents during the expedition. P. 6035 m was climbed on 19 July and P. 6055 m (their altimeter registered 6191 m at the summit) was climbed on 28 July.

(Article *HJ* 69)

P. 5700 m (Kange glacier)

Expedition / Organisers: French

Leader and (Members): Chance Emmanuel (3)

Period: September 2013

They began their trek from Akshow village and on 8 September established base camp at 3800 m near the snout of the Kange glacier. They established one more camp at 4500 m. From this camp the leader with Donadey Nicolas climbed the north face to reach the summit on 25 September in an alpine style push.

P. 5802 m and P. 5825 m (Pensilungpa glacier)

In September – October 2013, a British team from the Alpine Club led by Derek Buckle (leader) with Mike Pinney, Chris Storie and Tony Westcott visited the Pensilungpa glacier in the Zanskar region. In addition to general exploration of the area, they successfully made the first ascents of two summits on the Pensilungpa - Durung Drung divide; P. 5802 m (proposed name Hidden Peak) and P. 5825 m (proposed name Twin Peak).

(Note *HJ* 69)

Kishtwar Kailash (6451 m)

Mick Fowler and Paul Ramsden completed the first ascent of Kishtwar Kailash (6451 m) in alpine style from 4 to 10 October 2013 climbing 1500 m on the southwest face. They were supported by Mike Morrison and Rob Smith. This British team approached the mountain from Machail village and entered Dharlang nala. The base camp was established at 4000 m and after acclimatisation and reconnaissance; the peak was climbed in an alpine style push spread over a week.

(Article *HJ* 69)

Ladakh

Jungdung Kangri (6060 m)

James Monypenny (U.K.) and Cory Hall (Canada) made the first ascent of Jungdung Kangri (6060 m) that lies hidden in the Palzampiu valley of Ladakh. The team first tried to gain access from the south through

Likir, however, the pass could not be crossed by mules. Instead, they headed north by road to the Nubra valley. After a two-day walk with a couple of mules, the team reached base camp at 5090 m in a pristine alpine meadow. A high camp at 5522 m was established, below the west side of Jungdung Kangri. From there they reached the summit by a route which they named as 'the Monypenny Hall of Fame' (650 m, ED1 90° A2++) in an alpine style ascent.

P. 6046 m, Kang Yabat range

A small British team under the leadership of Douglas Briton climbed P. 6046 m in the Kang Yabat range in the Gya area of Ladakh. From Salsal on Manali - Leh highway, the team followed Yabat river to approach the mountain. After establishing three camps and a bivouac above base camp Caroline McCann and Matt Jones reached the summit via south face of the mountain. They proposed the name, Cha Ri, meaning a mountain looking like a large bird.

(Note *HJ* 69)

East Karakoram

Chamshen (7017 m), North Shukpa Kunchang glacier

An Indian British joint team led by Divyesh Muni and Victor Saunders, entered Sakang Lungpa and achieved first crossing of the Sakang col (6150 m) to reach North Shukpa Kunchang glacier. On the night of 14 August a wind blast of an avalanche from the slopes of Saser Kangri II struck the camp on the glacier and Andy Parkin suffered major injuries as he was thrown in a crevasse while he was sleeping in his tent. He was rescued by Indian Air Force helicopters in very unfavourable weather conditions. On 21 August, all remaining five members of the expedition with six supporting Sherpas reached the summit of Chamshen by the west ridge.

(Article *HJ* 69)

Plateau Peak (7300 m)

The Himalayan Club Kolkata section team, under the leadership of Debraj Dutta and consisting of eight members and 10 high altitude supporters achieved the first ascent of this giant of the East Karakoram. This peak was attempted many times during last few years by various Indian and joint expeditions without success. The Himalayan Club team approached the mountain from the western approaches via the Phukpoche glacier. After establishing three high camps, four members and five high altitude supporters reached the summit on 31 July 2013.

(Article *HJ* 69)

Rongdo valley explorations and climbs

An Indian Air Force team of 12 members under the leadership of Gp. Capt. V. K. Sashindran explored the Rongdo valley and climbed seven peaks during a 35 day long expedition in May-June 2013. This was only the third expedition in the valley. After establishing base camp at 4790 m, they established few high camps and climbed Sa'i Lhamo (6030 m), Khyung Kangri (6183 m), Chu Skeyes Kangri (6053 m), Langpoche (5968 m), Odgsal I (6234 m), Odgsal II (6028 m) and Charok (6123 m). They also attempted Lung Khor (6160 m) previously climbed and named as Gazgazri by a Canadian team in 2012.

(Article *HJ* 69)

(Full details and list of expeditions to the Indian Himalaya (2013) are available on the website of the Indian Mountaineering Foundation (www.indmount.org)

Other News

Dr. M.S. Gill, Past President of The Himalayan Club has had a long, abiding and continued interest in the Himalaya on a personal as well as professional level. A career bureaucrat, among other prestigious posts, he was Chief Election Commissioner who was awarded the Padma Vibhushan for his tireless work to bring in electronic voting machines and thus reduce malpractice. Later he was Minister for Sports and Youth Affairs for two terms in the Government. He is now a Member of the Rajya Sabha (MP).

Over all these years he has always been involved in the Club's activities – his six years of HC presidency was a landmark in that he played a very important role – that of raising funds to acquire a permanent home for the Himalayan Club – the first ever since 1928 when it was established.

As a fitting tribute, the Himalayan Club has conferred upon Dr. M.S. Gill the title of President Emeritus.

For the last 35 years at least, there has been a personality so closely associated with the Club, that fitting into his shoes is a marathon job for us editors. He has been editor of the Journal for over three decades, is the Human Wikipedia of the Indian Himalaya and has authored several books on areas and aspects of the Himalaya. An honorary member of most alpine clubs the world over, recipient of many honours, Harish continues to roam the mountains without a care.

If there is any worldly attachment that Harish has, it is to the *Himalayan Journal*. In recognition of his tireless contribution, the Himalayan Club has conferred upon Harish Kapadia, the title of Editor Emeritus.

The Second Jagdish C Nanavati Award for Excellence in Mountaineering was awarded to Debabrata Mukherjee for exploration over the Chaukhamba col, a high risk but wonderful maiden exploration of a very difficult route. This exploration linked the Bhagirath Kharak glacier with the Gangotri glacier, a most direct route. The Jury had made special mention of the Jadung Valley exploration by Ashutosh Mishra and his team, not only for a laudable exploration but also for a perfectly presented report that was to be used as a model example for applicants in the future.

The Garud Medal for outstanding support staff was awarded to a Darjeeling Sherpa, an unassuming man of few words but immense strength and pride in his work, Pemba Norbu, a.k.a King Kong.

Jim Perrin received the Kekoo Naoroji Award for his book *Shipton and Tilman* – a rare and wry insight into the lives of these iterant adventurers.

There were several deaths of friends of the Club. Aspi Moddie was more than that. He was an active family member, rather a father and a patron of the HC since 1949. He held posts of Vice President and President, he shifted the head quarters to Mumbai, broadened and widened membership, networked with international clubs - in fact he conducted overviews and financial projections long before the recent Strategic Review – exercises that held the Club in good stead for decades. He did this for 65 years! We at the Himalayan Club salute this great personality and mourn his passing.

Richard Martin Scott, our Honorary Secretary in the UK passed away in May 2013. One of the very few active overseas members, his loss is being felt very greatly.

The tragedy on Everest on April 18, 2014 when 16 Sherpas died while load ferrying and route opening, early in the season, left the world in shock. The business of climbing was temporarily shut down and Sherpas demanded better conditions and compensations. The debate on how to climb high mountains has begun again. Earlier, in the 2013 season, there were ugly fights, traffic jams on Everest and nine climbers were gunned down by the Taliban in the northern region of Pakistan.

The Himalayan Club Obituary

Name	**Class and Year of Election**
Richard Martin Scott	M 2005
Ian McNaught-Davis	H 2005
Nawang Topgay Sherpa	HC Tiger Sherpa
Michael Westmacott	H 1953

The Himalayan Club continues to move forward despite losses and setbacks. The year 2012 – 2013 and 2013 – 14 were eventful years for the Club in many ways. All sections were pro active in organizing interesting presentations. As always, Himalayan Club sponsored significant expeditions and climbers from the Club made great summits. Scholarships were awarded to many deserving candidates. Within the Club, the process of introspection and changes continued. A Strategic Review was conducted under Chairmanship of Vice-President Ravi Singh and the report, 'Scaling Up' was presented for consideration and implementation.

HC E Letters Volume 27 was presented in June 2013, Volume 28 January 2014 and Volume 29 July 2014. Membership is being streamlined and in recent times, social media forums such as Face Book are being used widely to promote the Club's activities among younger enthusiasts. Membership of the HC page on FB is over 20,000. The Himalayan Club website has also been revamped.

(Details of Club activities are available on the website www.himalayanclub.org and relevant E letters)

Full details and forms are available from Hon. Secretary, and on the Himalayan Club Website

THE HIMALAYAN CLUB CENTRE,

314, Turf Estate, Shakti Mills Lane off E. Moses Road, Mahalaxmi, Mumbai 400 011, India. Phone: 91-22 2491 2829 , E mail: info@himalayanclub.org, Website: www.himlayanclub.org

For Purchase of the *Himalayan Journal and Trade Inquiries contact*

Oxford University Press
67, C.S.T. Road, Vidyanagari PO, Kalina, Santacruz (E),
Mumbai 400 098
Phone: 91-22-2697 3891/92, 2652 1034/35, Fax: 91-22-2652 1133
www.oupin.com

Errata

Himalayan Journal **Volume 68.** Please note corrections as below:

1. Frozen Lake - Photo Feature by Sankar Sridhar (not Sankar Shridhar)
2. Photo Credits for the article 'The Irish in Himachal Pradesh'
   - Gramang Bal (page 109) - Sé Ó Hanlon
   - Kangla Tarbo and Shigri Parvat (page 111), Ramabang (page 112) and Bauli Khad col (page 113) by Gerry Galligan. (All four images were wrongly credited to Paddy O'Leary)

Printed in India by Parkar Arts, 1st floor, India Printing House, G.D. Ambekar Road, Wadala, Mumbai 400 031, Designed by Reflections and published by Nandini Purandare, Hon. Secretary for the Himalayan Club, 314, Turf Estate, Shakti Mills Lane, off E. Moses Road, Mahalaxmi, Mumbai 400 011, India.

THE HIMALAYAN CLUB

Climbing and Beyond...

THE HIMALAYAN JOURNAL

Volume 69
(2013 - 2014)

(incorporating the years 2012-2013)

Honorary Editors

RAJESH GADGIL

NANDINI PURANDARE

Honorary Business Editors

Rishad Naoroji

Shailesh Mahadevia

Designer

Aparna Joshi

Image by: DR. ANISH ANDHERIA

STAY SUMMER-PROOF WITH
CINTHOL
COOL SOAP
COOLING
CINTHOL
splash
CINTHOL
COOL
CINTHOL
ALIVE IS AWESOME

WORLD'S ICE DEPOSITORY IS MELTING
DUE TO
GLOBAL WARMING
VANISHING GLACIERS GIVE AN ALARM NOTE FOR MOTHER EARTH.

Melting of Himalayan Glaciers

PLASTICS ARE AMONG THE HIGHEST GREEN HOUSE GAS SAVERS
&
DECREASE THE IMPACT OF GLOBAL WARMING

DON'T LITTER, USE PLASTICS RESPONSIBLY, KEEP THE MOUNTAINS CLEAN

Plastics, Metals, Paper ...
Can be recycled into useful products.

PLASTICS ARE 100% RECYCLABLE

Issued in Public Interest by ICPE

www.nanhikali.org

This is 5 year old Shreya. Her parents abandoned her and vanished. If you turn the page, you'll do the same.

With no one left in the world, Shreya doesn't have the slightest hope of facing the future, unless she gets an education. And that's where you come in, to make a dream called Nanhi Kali come true. Through Nanhi Kali you can become the foster parent of a little girl like Shreya and support her education.

All you need to do is contribute Rs. 1800/- per annum and your 'daughter' will get to go to school. We will send you regular reports on her progress, including a photograph.

So please don't turn the page yet. Come, join the Nanhi Kali movement and together let's ensure that every girl child in the country completes ten years of formal education.

Nanhi kali
for the girl child

A project jointly managed by
K. C. Mahindra Education Trust and **Naandi Foundation**

Space donated by Mahindra & Mahindra Ltd.

Yes I,________________would like to adopt a 'Nanhi Kali'. Enclosed is my cheque of Rs.1800/-, in favour of K. C. Mahindra Education Trust A/C Nanhi Kali, for one year's education of one Nanhi Kali or I would like to adopt ________ Nanhi Kalis for ________ years. My cheque for Rs. __________ is enclosed @ Rs.1800/- per Nanhi Kali per year.

Please send your cheque to: K.C. Mahindra Education Trust, Cecil Court, 3rd floor, Near Regal Cinema, Mahakavi Bhushan Marg, Mumbai - 400 001. Tel.: (022) 22021031. Fax: (022) 22852441.

All donations are exempted under Section 80 G of the Income Tax Act.

Pidilite
Pidifine™
50 YEARS
of experience in pigment dispersions...
Now for paper
PIDIFINE 100 SERIES
PAPER COATING & PRINTING
• Wide range of brilliant colors
• Compatibility in coating formulation
• Optimum particle size for brilliant effects
• High fastness properties
• Easy runnability
PIDIFINE 500 SERIES
PULP COLORATION
• Excellent light fastness
• Easy on effluent load
• Good stability under high speed running
Pidilite Pidilite Industries Limited
PO Box No.17412, Ramkrishna Mandir Road, Andheri (E), Mumbai-400 059.
T: (9122) 2835 7510/7220/7000 F: (9122) 2835 7008 E: vykokane@pidilite.com W: www.pidilite.com

A TATA Enterprise
EXCLUSIVELY AT
WESTSIDE
/WestsideStores
/westsidefanpage

INDEX

TO

THE HIMALAYAN JOURNAL

VOLUME 69

THE HIMALAYAN CLUB
Climbing and Beyond...

THE HIMALAYAN CLUB
314, Turf Estate, Shakti Mills Lane,
Off E. Moses Road, Mahalaxmi, Mumbai 400 011, India.
E-mail: info@himalayanclub.org
Website: www.himalayanclub.org
Phone: (+91-22) 2491 2829

E mails

Club Office	:	info@himalayanclub.org
Hon. Secretary	:	secretary@himalayanclub.org
Hon. Editor	:	editor@himalayanclub.org

Note on the Index

Articles/Expeditions and Notes : listed under the author's name and under the name of the peak and region.

Obituaries : indexed under the name of the deceased.

Book reviews : indexed under name of author.

Maps and sketches : indexed separately.

The names of book reviewers and authors of obituaries have not been listed.

Index to
The Himalayan Journal
Volume 69

Compiled by : Nandini Purandare

Index of Maps and Sketches

Printed at Parkar Arts, India Printing House, 42, G.D. Ambekar Marg, Wadala, Mumbai 400 031